MW00681598

What Great Telecom Managers Know

How to get the recognition you deserve by demonstrating your value and ROI

Roger K. Yang

Published by:
Simcoe Publishing
1055 Bay Street, Suite 1110
Toronto, ON M5S 3A3

ISBN 0-9738138-0-6

Printed in Canada

Acknowledgments

Every quality book is a tremendous effort, not just by the author, but also by all the many people who helped along the way. This book is no different. I don't think I would have completed it if it weren't for all the help I received from others.

The purpose of this page is to formally say "thank you" to everyone who had a hand in putting this book together. Most importantly, the basis of this project is all the telecom professionals who I've met over the years whose experience and knowledge I learned so much from. A special thank you goes out to all the telecom managers who agreed to spend some time being interviewed, and those who attended my workshops.

Also, I appreciate the subject matter experts whose opinions influenced much of the content. Just as important, several Avema employees and partners contributed many ideas and diligently read and re-read my ramblings. Below are a few of these people:

Anne Cateaux   Vasilis Dermosoniadis
Carl Friesen   Ivan Glisin
Jonathan Gullery   Anita Harnarain
Chad Hughes   Brian O'Donoghue
Scott Salsbury   Willa Schecter
Larry Treas

Table of Contents

Section I

What is a Telecom Professional Worth?

Chapter 1

A Day in the Life
of a Telecom Manager

*In what ways would you like to benefit
from reading this book?*

What the Author Knows
About Great Telecom Managers

I am not a telecom manager myself, and I write this as an observer with an impartial point of view. Viewing the function of telecom management outside of your day-to-day routine, from someone else's eyes, may help you to gain more understanding to propel your career even further.

Since 1995, I have worked with hundreds of telecom managers to help them manage their costs and streamline processes. During that time, I have been honored to learn much from all these great people – and I hope they have learned something from me as well.

The inspiration for this book came throughout 2004, while conducting seminars with telecom managers in various countries. I've met with telecom professionals throughout the United States, Canada, the United Kingdom, Ireland, and Greece, to name a few. Through these seminars and meetings, I've come to realize that despite the different tariffs, telcos, standards and markets, the issues faced by telecom managers are universal around the globe.

I realized that there could be tremendous value in compiling this understanding, gleaned from so many different sources, into a single resource.

This book does not focus on the technical "nuts and bolts" about PBX configuration, or the nitty-gritty details about phone bill analysis. There are many other resources available in those areas. Instead, this book is specifically about how telecom managers can be a more strategic part of their organizations, maximizing their value and increasing their career satisfaction (and, hopefully, their salaries).

To gain a deeper insight about telecom managers, their frustrations, challenges, and motivations, Avema Corporation has conducted a comprehensive survey.

For this, we developed a detailed survey and then interviewed telecom managers by phone throughout the United States, Canada, and other countries around the globe.

This was not just a superficial five-minute, online questionnaire. We thought it worthwhile to spend the time on the phone with these telecom managers, because only by doing this could we ask probing questions that gave us a better idea of what real-world issues they were facing. We were deeply grateful that these busy people each took the time from their schedules to give us a glimpse into the world of their roles within their organization.

While some of their answers confirmed what we already knew, we found it helpful to get a wider opinion on current issues. The fact that much of the information in this book is based on the results of the survey gives you, as a reader and telecom manager, the confidence that you're getting wisdom that has been tested out in the real world.

How to Maximize Your Benefit From Great Telecom Managers

As I was writing this book, I thought of changing the title to "What Great Telecom Managers *Do*" instead of "What Great Telecom Managers *Know*," because all the difference in the world comes from *doing*. But my editor didn't think the title sounded as catchy. To get the most out of this book, it takes more than just reading. In fact, anything worth doing well takes more than just passive reading. Comments that I get from participants in my workshops suggest that they can pick up a lot of knowledge, but time needs to be taken to apply that knowledge, and make it a part of everyday life.

The company I work for, Avema Corporation, has a mandate to help telecom managers be as successful as possible, since our business is all about helping telecom managers improve their work environments through what we call software "power tools." In that same spirit, this book is all about helping to make you as successful as you can be in your career.

Some of what you will read in this book may give you new ideas, and some other parts will reinforce what you already know. I wish you all the success you desire throughout your career.

Telecommunications Is A Critical Part Of Every Organization

Anyone who reads this book knows how important the telephone systems are to a large organization. Without telephones, data connections, and mobile phones, every corporation, government, university and hospital would come to a grinding halt. The entire economy, and by extension, our modern society, absolutely depends on these telecommunications systems to be working at all times.

Businesspeople say that cashflow is the lifeblood of a company. If that's true, then telecommunications are all the connections inside the brain and the rest of the nervous system that allows the body to function.

So why is it that some executives place so little value on the telecom department?

As a telecom manager, do you sometimes feel taken for granted? You're not alone. Many of your peers feel the same way. In many cases, this isn't just a feeling. Most other people in an organization do not understand the complexity of managing telecom equipment, services, vendors and invoices. They take for granted that when they pick up their desk phone's receiver, they get a dial tone. If they move to a new office, they expect that their former DID or extension will be transferred to their new location, exactly the same as when they carry their desk organizer and picture of their family over to the new desk.

Mostly, people do not think about telecommunications until something goes wrong. Of course, when that does happen, it's assumed that it will all be fixed by flicking a switch.

Anyone who manages telecom knows that it's not that easy.

Theoretically, the organization could outsource the telecom function and rely on outside vendors to manage all its communications equipment and services, but the response time could be longer and costs could quickly get out of control, dramatically affecting the bottom line.

How can you make the importance of your role more visible? How can you demonstrate your strategic business value to the organization?

Salaries and Work

The Voice Report, a newsletter focused specifically on telecom managers, publishes an annual report on telecom professionals' salaries. Here are some of the findings from their 2004 National Salary Survey:

> *"It's been a rough year for voice pros, who find themselves facing more job responsibilities but no matching boost in their pay packets."*
>
> *"More and more work is being heaped on telecom managers' shoulders, ranging from VoIP projects to wireless and video management. As a result, voice pros are finding themselves putting in increasingly longer hours under mounting levels of stress."*

40% of the telecom managers surveyed felt that their salaries were unfair. Despite this less than ideal pay and all the hard work, 78% indicated that they are satisfied with their jobs.

It's obvious from these statistics that telecom managers are a dedicated bunch who generally enjoy their work, even if it means that they get less recognition, and often less pay. The Avema survey supports these findings with similar conclusions.

Chapter 2

How Everyone Else
Views Your Job

Think of how the head of your organization views the telecom function. What does this person want from you as telecom manager? What more could you do to meet his or her expectations?

If You're Not Demonstrating ROI, Your Salary Is <u>Just Another Expense</u>

Even though you may do what you can to advance your career and job satisfaction, a big part of your overall quality of work life depends on factors outside your control.

To deal with this factor, it helps to bear in mind the classic, and still relevant words of Dale Carnegie, the master of human relations, who wrote *How to Win Friends and Influence People* in the 1930's, *"Talk in terms of the other man's interests."* Of course, today we would update that to "other *person's* interests."

A more recent quote from *The One Minute Sales Person* (Spencer Johnson and Larry Wilson, 1984): *"I have more fun and enjoy more financial success when I stop trying to get what I want and start helping other people get what they want."*

So, on whose needs should you focus, if not your own? Who is the person who likely controls your salary more than any other? In today's economy, it's typically the head of Finance in your organization.

This person's responsibility is to make sure that the organization has the financial resources it needs to do its work. Sometimes this means getting outside funding, possibly through a share offering, bonds, grants or donations.

But it also involves watching the expenses of the organization. It's the Finance department's role to make sure that the top line of the organization's income statement – the revenue – doesn't get diminished too much by the time all its expenses are subtracted, hopefully leaving a sizeable profit.

Where do we find Telecom in the world of the Finance department? Unfortunately, it is typically slotted under "expenses."

And that's a problem. In many organizations, telecom costs come just behind payroll and rent. That is a BIG line item to control, and a good deal of the worry that goes into each day for the CFO and controller, whose jobs revolve around keeping costs down. Finance may be pleased to see long-distance costs plummeting, but in many cases these are counterbalanced by mushrooming costs for wireless.

Of course, you still need to perform all the technical work, MACs, updates, and generally keep everything running. Unfortunately, you won't get noticed for maintaining 99.999% uptime (but you will be noticed if the communications networks go down, just not the way you'd like), because that's expected of your department anyway. Driving more ROI in a visible way will get you noticed much more.

In my experience, it's the telecom managers who understand the needs of their organization – particularly those of the Finance department – and help to meet them, that have the highest levels of job satisfaction.

Defining Telecom Department ROI

Every finance manager's career lives and dies by ROI, "Return on Investment." You are probably familiar with the term when used to describe a project, or when purchasing new equipment. Have you ever thought of ROI as it relates to your own job function? What is your telecom department ROI?

The "Investment" part of the equation refers to your department's salaries and associated costs like health benefits and office space, as well as the tools that you use, such as your computer and software. Basically, whatever expenses are associated with managing the equipment, services, vendors, and invoices fall into this category. This does not include the actual cost of the telecom equipment and services.

"Return" is any positive impact that your department has on the organization's bottom line. Cost savings from lowered prices go directly to the bottom line, as well as savings from any reduction in telecom use. Anything that you can do to enhance the organization's sales or revenue is also categorized in this area.

Most of the time, the telecom department will produce ROI through cost savings. Although this may not seem as glamorous as generating sales, every $1 in cost savings is far more valuable than $1 in new revenue. Revenue is a gross figure, and the costs associated with generating it like labor and raw materials must be subtracted. For many companies, it may take $10 in revenue to earn $1 in profit. In contrast, every dollar of cost savings directly affects profitability by a dollar.

The values of publicly traded companies are measured by

their profitability, often expressed as a Price to Earnings (P/E) ratio. A P/E ratio of 20 means that for every $1 in profit that you generate from cost savings, the value of your company increases by $20!

The formula for calculating your telecom department's ROI is very simple. If the calculated return totals more than the investment, you have a positive ROI. If both the return and investment are about equal, the ROI is neutral. If the return is less than the investment, then you may have reason to worry.

The only two ways to increase ROI are to either increase the return, or lower the Investment. In Section II of this book, we'll look at some tactics that telecom managers can take to increase the "R." In Section III, we'll look at the software tools that can help your department do more with fewer resources, in other words, helping your department on the "I" side of the equation.

ROI in the 21st Century

A few years ago, the CIO was almost as important as the CFO, and had much more discretionary spending authority. At that time, the technology function was considered strategically important, and it was assumed that there was tremendous ROI that would come naturally. Also, executives believed that massive spending on technology was necessary to keep up with competitors, to prepare for Y2K, and to get in on the "new economy."

But ever since the excessive spending and letdown from Y2K and the dot-com bubble, companies have been much tighter on IT spending. Billions of dollars have gone out the door, and not enough questions were asked about how much the benefit was coming back to the organization. Since then, most companies now have the CIO reporting to the CFO, and IT's budget is much tighter than before. IT and telecom costs are being scrutinized, and executives are looking for every way to shave pennies.

Telecom, as many readers will know, often reports to the CIO, and since CIOs do not usually have a telecom background, they prefer to focus on things like desktop computers, servers, and data networks.

Today, CIOs are just as likely to be focused on ROI as CFOs. This is all the more reason that telecom managers need to focus more than ever on their own ROI.

Remember:

"If you're not demonstrating ROI, your salary is just another expense"

Chapter 3

Telecom Manager Career Case Studies

*What can you learn from reading about
other telecom departments?*

From all the many telecom managers that I've spoken to, two in particular stood out as being classic "best case, worst case" examples. I've gone into these in some detail – one as a warning, the other as an example of how a telecom department can change and evolve.

In the first case, the name of the individual and some details of the case have been changed to protect this person's privacy. The second case involves a major retailer.

Case One: The Challenges of a Call Center

"Why can't a phone bill walk through the door and be right?" is the heartfelt request from "Marilyn," the telecom manager in a US-based call center. The need to check phone bills and get carriers to correct them is one of the biggest parts of her job.

During her 35 years in the business, Marilyn has experienced many changes. This includes seeing equipment shrink dramatically – a given volume of traffic might have required four floors' worth of equipment back when she started out in telecom, and can now be carried on three racks.

But those error-prone phone bills are still with her.

This problem gets larger, not smaller, as her company mushrooms with the growing business of collecting debts. When she started with her present employer two years ago there were 350 employees. There are now 1000, and there are plans to soon add many more.

While her workload has tripled over the past year, she is frustrated that her budget has only doubled.

Marilyn is also frustrated by the shortcomings in how her personal and department performance are measured. The goals she must meet are set by senior management and are outside her control.

In many cases, she is not given the tools she needs to achieve the goals as set. One issue is training – there's not enough of it. Whenever money is short, as it often seems to be, "training always gets zapped out of the budget, and so we spend ten times as long learning how to do something than we would have with proper training."

As well, the goals she must meet don't stay put. "They change every day," she says. "It's all a moving target, it's management by crystal ball."

One of the biggest variables in her work life – the external costs – is not something she is able to control.

She does what she can to pressure carriers and vendors to lower their prices, and she consolidates operations to avoid duplication. Sometimes the company's penny-pinching has exposed them to risk, she says – disaster recovery is not getting nearly the management or financial attention it needs.

The good parts to her work include the fact that her department's contributions are valued "once in a while." There's an occasional pat on the back when things go well. And, she's certainly never bored; the work always changes.

Marilyn knows her work is vital to the company's success, but she feels her salary is not in keeping with her value.

Although dealing with day-to-day issues takes up much of her time, Marilyn works at staying up to date. She gets support from the users' group of her telecom equipment vendor, and finds their annual conference worthwhile. She also reads call center and telecom-related publications.

Looking to the future, she doesn't think that VoIP will simplify her work much or help reduce costs. She hasn't seen much convergence in her work yet. There's not much room for VoIP in her company, because much of its work is done on government contract and this requires the use of standard telephony. "This holds us back," she says.

Marilyn believes that mergers in the telecom industry will not make life easier – larger companies will mean that the billing accuracy problem will just get larger.

Her plans include "cradle-to-grave" call records, important particularly in collections work. It is helpful to be able to follow a call from start to final hangup, including any transfers and on-holds. Perhaps most important is to know whether it was the agent or call recipient who ended the call – often, the debtor who received the "reminder" phone call from a call center agent will claim that it was the agent who hung up, while good records would show that it was in fact the recipient who hung up on the agent.

Although Marilyn is a diligent and knowledgeable worker, and doesn't dislike her job function, her situation certainly isn't ideal.

Case Two: The Bright Side at a Retailer

It's a night-and-day comparison between the circumstances in the northeastern call center and the situation faced by Cathy Halton at a major retail chain with operations across the United States. Cathy has 19 years of experience in telecom, and is now Networking Engineer.

Cathy says that when she started with the company about five years ago, Management's attitude towards telecom was, "A phone is a phone. Big deal. Make them work."

Back then, she was the third person hired in the telecom department; they now have 12, and getting more staff is becoming increasingly easy to justify. In the early years, telecom was considered a cost center and any funding for projects such as upgrades had to come from their internal revenue. Now, it's easier to get funding.

What made the difference? Cathy says that it is partly due to it being a forward-thinking company in growth mode, with Management who will listen to new ideas. But it was largely due to some extra-hard work by Cathy and her team.

Their first task was to make the current system run smoothly and cost-effectively. Doing this involved contract negotiations and consolidating dial tone. Once this was accomplished, they were able to do a presentation to Management, showing how telecom costs per location were going down, even as the number of locations was growing.

While working hard on this, they devoted after-hours time to developing some new projects that would add value and make Management take notice. Cathy cautions anyone wanting to move from what might be called a back-office manager into something more strategic, that they be prepared to put in the time needed to make it happen. In her case, it was "a few hours per night for a few months."

However, once "over that hump," as she says, Management started to look at the telecom department as more than just a necessary expense. Now, they get to work with other departments to roll out exciting new projects, during normal working hours. In other words, Cathy was able to demonstrate a significant Return on the company's Investment in the department. This began a cycle where the company is willing to put even more into its Investment, and Cathy's department is able to generate even more Return.

One of these has been meeting the request to provide electronic photo-printing services, where customers would use any Internet-connected computer to upload their digital photos to a third-party server, select the location where they wanted to pick up their prints, and the prints would be made available there. Setting this up required a large investment in bandwidth and the ability to deal with security concerns.

Success in projects like this mean that now, 80 percent of Cathy's time goes to strategic projects.

Cathy is also pleased with the way her efforts are evaluated. While the overall goals for the company are set by senior Management, she works with her own manager to set her goals and those of her department. It is important that these be "SMART," she says, meaning "Specific, Measurable, Attainable, Reportable and Timely."

Some of these goals relate to savings – "We will save the company $X on telecom this year." Others are more project-based, and there is a rigorous process that involves setting the goals for each project, such as expected revenue and costs, and reporting periodically on any variances.

For the future, she wants to develop a better way to manage projects. Success breeds the expectation for more success, and now she finds other department managers coming to her with requests that she undertake tasks for them.

She finds that scheduling these requests is important – it does not work well to just say, "Sounds like a good idea. We'll put it on our list of projects, and get back to you." Rather, she needs better project-management tools so she can get a clear idea of when her department's time will be freed up enough to tackle the new project. This is important when she gets urgent requests or even demands from some management people who insist that she implement their favored project, right now.

Cathy would also like to invest in advanced software to reduce the amount of time her department spends on administrative work, freeing up even more time for higher value strategic activities. It would also help the company meet the rigorous demands of the Sarbanes-Oxley Act of 2002, which pushes organizations to provide better documentation of all their work. Any way to automate the tasks of the company, and show a convincing paper trail of documentation, helps the company meet the legislation's requirements.

Clearly, Cathy enjoys what she does and enjoys being acknowledged for her importance to the company. "I love my job," she says, "because there's value."

What Can We Learn From These Case Studies?

One lesson to be learned from these studies is that just because telecom is vital to the organization's functioning does not necessarily mean an easy, rewarding life for the people who make sure that the telecom function happens. We can see this in the case of the call center – making and receiving phone calls is certainly the most vital function of all for this organization. If the phones don't work, the employees cannot do their jobs, and revenue drops to nothing. If telecom costs are not controlled, profitability may vanish. However, it seems clear that at least in this call center, telecom is treated as a low priority, just a necessary cost of doing business, rather than a strategic resource.

At the retail company where Cathy Halton is employed, one might argue that telecom is peripheral to operations – the key strategic functions of a retailer are buying the right merchandise and the logistics involved in getting it onto the shelves. Even though telecom is a support function, Cathy is clearly given an important role to play in the company's future. In fact, her telecom skills are integral to helping another profit center – the digital photography project – to blossom.

It's quite possible that both companies are financially successful. The call center is definitely in growth mode, as the ever-growing personal debt of Americans is fertile ground for any company able to help collect some of that debt. The retailer is in growth mode as well. However, at the call center the extra volume just means more work for the telecom manager, while for Cathy it means more resources (staff, dollars) for the telecom department.

The retailer has forward-thinking management. Perhaps the night-and-day difference between these two situations is due largely to the company's willingness to listen to new ideas and invest in those that show the most promise.

Even then, it wasn't easy. As Cathy says, getting "over the hump" from back office function to strategic advantage took imagination, tenacity and a willingness to put in substantial extra time on the job. Not all telecom managers are in a position to do this, or want to.

Chapter 4

Best Practices of Telecom Managers

How can you use these best practices to improve your own career, and how will you make that happen?

In writing this book, one of my main goals was to help telecom managers learn from each other. There is a tremendous wealth of knowledge and experience within every telecom manager, and connecting this all together makes for a very useful resource.

To understand some of the best practices, let's take a look at the common traits of telecom managers who report the highest levels of job satisfaction.

Strategic Vision

Our experience is that top telecom managers are very conscious of their ROI, and make sure that their managers are also aware of it. They are constantly seeking ways to deliver more results, and spend a good portion of their time on planning for the future. Although almost a third of telecom managers measure ROI for projects and purchases, only about 10% measure their own department's effectiveness in terms of ROI.

Those who report the most job satisfaction are more likely to spend the majority of their time on strategic work, rather than day-to-day tasks. When I speak with these telecom managers, I can sense strong feelings of pride in their workmanship. They are very animated about their plans for the future, and they very much enjoy what they do.

Setting Goals

An important part of success in telecom management, as in any endeavor, is to set goals, write them down, and focus on your target. We work far more effectively when we have strong goals that we are committed to and that we believe in. Chances are, you've heard a lot about goal setting from various self-help gurus, and in the general media. What's really interesting about the whole concept of goal setting and achieving, is that if asked, most people will agree that it is helpful, yet very few people actually do much about it in their personal or professional lives.

Why does goal setting work? Imagine being given a bow and arrow and pointed towards a target. Chances are, you will be able to get the arrow at least in the vicinity of the target. Now, imagine being blindfolded, spun around, and asked to shoot at the same target. Pretty hard to do, isn't it? How can you reach a goal, if you don't even know where it is? Take the blindfold off. The more you focus on that target, the more you practice hitting it, your odds of success increase dramatically.

As human beings, we pay more attention to things that we think more about. Have you ever been shopping for a car, spent some time deciding on a specific model and color, and the more it's on your mind, the more you notice many of the same model and color car in parking lots, or driving by? Or if you just bought a new piece of clothing, you start to notice other people wearing either something similar, or the exact same thing?

When you drive, or if you ski, instructors tell you to look towards where you're going, because you will tend to follow wherever your head and eyes are pointed. If you focus on the road ahead of you, you'll continue to go straight, however, if you look over at the car next to you, or the landscape off to the side, you will start to veer towards it.

By setting clear goals, focusing on what you desire as an outcome, and spending more time thinking about it, you are far more likely to reach those goals. Just like the examples with the car and clothing, your mind will more likely pick up pieces of useful information or ideas related to those goals.

As we saw in the second case study, there are some specific aspects to goal-setting that make it work.

Goals are ideally set by both the employee and the employer, so that there is an agreed upon metric of how "success" is defined. If your employer isn't progressive enough to work with you on this, then it's still a worthwhile exercise to do for yourself.

As Cathy Halton says, goals must be SMART — meaning "Specific, Measurable, Attainable, Reportable and Timely."

Writing your goals is a very effective and proven strategy, because it forces you to dedicate some time to focusing on specifics, and about what your success will be like when you achieve it. The action of picking up a pen and writing your goals also helps to solidify your commitment to the outcome. It also helps to make them public. This is one of the reasons for the success of group-based change-management programs such as Weight Watchers. All of these factors will increase your level of success dramatically, if properly applied. What is most important is your own internal dedication to success.

If you're serious about achieving top results, do not read any further in this book until you spend some time to write down your specific goals. Only continue reading once you have clearly lined out what you want to accomplish, as well as a plan to achieve your goals.

Reporting Your Results

As you achieve success, you want to make sure that other people hear about it! Even if you are actively achieving a tremendous ROI, if your managers and your managers' managers aren't acutely aware of it, you will still be viewed as just another expense.

Maryann from the case studies in the previous chapter is a great example of reporting achievements. The telecom department is very proactive in reporting their results, and because of this, the senior executives in the bank consider telecom savings important enough that they include them in their annual report. Imagine, this information is broadcast to every shareholder and employee, who can see the results from the telecom department's work.

How do you report your results? Speak the language people want to hear. In some organizations, this is best done through written reports. In others, it's through colorful charts; in others, it's verbal presentations. Some people may do best with tables of numbers.

In addition to reporting your results on an occasional basis, it helps to keep your end users involved with new projects or installations through training, regular communication, and maybe a website or intranet page that gives more useful information.

Working On Your Organization's Objectives

Management doesn't recognize what you're doing unless you tell them. Communicating your value, as we discussed above, is part of the solution.

But it needs to be a two-way street. Looking back to our "Best Case" scenario at the retailer, we see how Cathy Halton worked with the other departments to develop a solution to a market opportunity. You need to do likewise – find out what the organization's needs are, and then see how you can become part of the solution.

This means talking with other people in the company, and also reading the trade press not just of the telecom industry, but of your organization's industry as well. Are you in a retail organization? Government? Healthcare? Hospitality? If you aren't already reading up on these publications, make sure to find out where you can access them, and spend some time keeping up to date on industry trends. Chances are, many others throughout your organization already subscribe to these publications.

Networking with your industry peers is a great way to share information about upcoming trends. Organizations such as the ACUTA, the Association for Communications Technology Professionals in Higher Education (www.acuta.org), and NASTD, Technology Professionals Serving State Government (www.nastd.org), are examples of organizations focusing on telecom professionals within an industry.

Continuous Improvement

The top professionals in any field are constantly looking to improve on their knowledge, skills, and success. The fact that you're reading this book to upgrade your knowledge is one example of that.

If you think of any top athlete, they all have coaches. Michael Jordan was arguably the best basketball player, far beyond anyone else at his time, and he had coaches to continuously help him to improve even more.

Every business professional at the top of their game constantly reads books, magazines, and attends seminars and workshops to improve their knowledge and their success. Even if they only gain a small edge from each of these endeavors, they consider it a worthwhile exercise.

Here are some resources that are widely used by telecom managers in North America:

Publications

Many of these publications have both print and electronic versions available. Some people prefer the feel of a printed newsletter, while others prefer to read online. Personally, I find that my time is best utilized by subscribing to electronic formats, with summaries sent regularly by e-mail.

The Voice Report is an interesting publication to note, not only because it is written specifically for telecom managers, but also because the content is frank, and designed to be practical and useful. It is more expensive than a typical magazine or trade publication, but that's because it doesn't contain any advertising, and isn't influenced by any vendors.

Here is a list of some telecom-related publications that your peers are reading:

- CommWeb (www.commweb.com)
- Computer World (www.computerworld.com)

- Call Center Magazine
  (www.callcentermagazine.com)
- Business Communications Review (www.bcr.com)
- The Telecom Manager's Voice Report
  (www.thevoicereport.com)
- Information Week (www.informationweek.com)
- Network World (www.nwfusion.com)
- Telemanagement (Canadian)
  (www.decima.com/reports)

Organizations

These are the organizations that telecom managers find most useful in their careers, according to our survey:

- International Nortel Network Users Association
  (www.innua.org)
- International Alliance of Avaya Users, Inc.
  (www.inaau.org)
- Joint Users of Siemens Technologies (USA)
  (www.just-us.org)
- Cisco IP Telephony Users Group (www.ciptug.org)
- Association of Telecom Management Professionals
  (www.aotmp.com)
- Government Management Information Sciences
  (www.gmis.org)
- NASTD Technology Professionals Serving State
  Government (www.nastd.org)
- ACUTA Association for Communications
  Technology Professionals in Higher Education
  (www.acuta.org)

Peer Networking

There are numerous conferences that are run by the organizations and publishers that I just listed. INNUA, for example, has an annual Global Connect event where they regularly have 3,000 attendees. They also have regional meetings in the United States, where the atmosphere is a bit more relaxed since the crowd is much smaller, at about 250 attendees. And there are local chapters that meet on a regular basis. These smaller gatherings are better for developing closer ties with your peers, although the larger conferences certainly generate more excitement.

In addition to these conferences, there are a few e-mail discussion groups where telecom managers can post a question that they've been struggling with, and get responses from numerous helpful individuals within hours, or even minutes. I subscribe to Telecom-Talk, which is run by The Telecom Manager's Voice Report, and it is a great resource. Here are some e-mail discussion groups:

- Telecom-Talk
  (www.thevoicereport.com/listserv.html)
- Terry Grace's lists, including Cisco Call Manager and Nortel user discussions
  (www.tgrace.com/mailman/listinfo)
- Association of Telecom Management Professionals
  (www.aotmp.com available through membership)

Training

Below are some of the larger seminar and training organizations that focus on telecommunications:

- Global Knowledge (www.globalknowledge.com)
- Tera Training (www.teratraining.com)
- CCMI (www.ccmi.com)
- Business Communications Review (www.bcr.com/training)
- Telemanage Training (www.telemanagetraining.com)

Mentorship

A mentor is someone, generally more experienced than you, who provides advice and connections along the way through your career. Most do this out of a simple motivation of generosity, partly because someone else helped them. And also, many people like to hand out advice; it's in their nature.

Because these are informal relations, there's no central source or a website where you can find mentors. Start with senior people in telecom, although you may find a good mentor in almost any other role. Some with management experience in another organization or another role, for example, may be able to help you deal with management issues. You may already have a mentor-like manager that you report to, as some talented senior managers often help their subordinates to be as successful as they can. If not, it may not be best to cast the person you report to in a mentorship role, as it could possibly put them in a position of conflict of interest, and they may not be able to be as disinterested as they should.

Coaching

A business coach is something like a mentor, but it's a professional relationship in that this person is paid for her/his time. Your HR department may be able to find you a good coach or you can find one through a local coaching organization. You may start with the International Coach Federation, www.coachfederation.org.

A coach will, through appointments that are typically a half-hour long, generally by phone, help you understand the issues you are facing. The coach does this mostly through asking you questions that help you find your answers for yourself. Only rarely will the coach "tell" you what to do – the expectation is that you already know the answer from your previous experiences – you just need help uncovering it for yourself.

You should interview several coaches to find one you think has what it takes. There needs to be good personal chemistry in the relationship, and you need to be confident that the coach you choose will take an approach with which you feel comfortable.

Certification

Although there are vendor-specific technical certification programs for various equipment, there has not been a certification program for telecom management as a profession until recently. The Association of Telecom Management Professionals has been offering a Certified Telecom Management Specialist program since the summer of 2004 to fill this need in the marketplace.

This program focuses on cost management, inventory, invoice validation, and contract management, and includes 5 days of training and an exam.

Job Satisfaction

A major purpose of conducting the Avema survey was to gauge telecom professionals' job satisfaction and some of the details that are involved. Since the survey was done on a volunteer basis, we can assume that those who participated are more likely to be more passionate about their careers in telecom, and likely more progressive than most telecom managers. Here are some of the key findings:

Recognition

We've already established that the telecom department is not very well understood, particularly voice. It's also obvious that telecom managers don't get the recognition that they deserve for keeping this critical infrastructure running. Since phones always work, it is assumed that it's no trouble to keep them going. Everybody has a phone at home, and one on their desk. Every time you pick it up, it works!

Computers, by comparison, are so much more complicated. They crash, they give us arcane error messages, and a typical user doesn't know how to deal with it effectively, so we understand that we need "experts" to manage them. Many IT troubleshooters are widely hailed as heroes for their ability to debug a balky network, recover data lost in a hard-drive crash, and seek and destroy destructive worms and viruses.

Over the years, people studying management and motivation have repeatedly witnessed that what motivates employees is not so much money, perks, or even a window office. What motivates them the most is recognition from their boss and their peers. We see this in military situations where combatants will enter potentially lethal circumstances willingly and even eagerly, and are rewarded with the honor of serving their country, symbolized by a medal on their lapel.

This kind of recognition is much more important than it may seem. Despite all the management training, courses, and publications that demonstrate how important it is to employees, most managers do not use it very effectively.

This message comes through loud and clear – telecom staff aren't getting the recognition they need. They know how much effort and skill it takes to assure the high level of service expected of telecom services. They know the effort it takes to make sure that MACs happen smoothly and that all of this happens within their budget.

Career-boosting suggestion: "Charity begins at home" is a well-established saying. It could be that "recognition starts at home" too. So, if you have staff members who report to you, consider how often you provide recognition to them for a job well done, going above and beyond the call of duty. Most managers, even if they are fully aware of how important recognition is to their staff, rarely practice it. If you feel starved for recognition yourself, can you understand what a sincere gesture might mean for the people who report to you?

Consider one of your technicians, who pulls an all-nighter to set up the system in a new office, and then comes in bleary-eyed (and only a little bit late) for another full day of work. Did you write a letter of commendation for the tech on behalf of your boss, and then ask the boss to deliver the letter to the tech in person? Was the technician recognized for her/his contribution at your next staff meeting? Was there a "thank you" or an acknowledgement of a job well done, in the form of a dinner-for-two certificate to a prestigious local restaurant?

Giving recognition to your staff doesn't have to be expensive, but it can certainly be rewarding to the staff member –

and to you. It makes the person to whom you report realize just what lengths the telecom department will go in order to help the organization do its work. You get recognized for having pulled together and grown an effective team.

Most and Least Liked Parts of the Job

A big part of job satisfaction, of course, is having a clear idea of what you like and don't like. If you think through the various facets of your job, you may find that there's a whole lot more to like than dislike.

But what do other people say? In asking what were the favorite and least favorite parts of the job, we discovered a surprising amount of agreement – and also disagreement. Most telecom managers very much enjoy their work, and are quick to explain what it is that drives them.

By far, the most enjoyed aspect of the job is the variety. About half of those surveyed commented that they enjoy learning about new technologies, and the challenge of adapting to all the changes and different tasks throughout the workday. The pace at which these tasks must be accomplished is also something that many are very comfortable in dealing with, and some mentioned that they would be disappointed if it slowed down.

Many telecom managers enjoy finding solutions to help other people in their organization. They have a strong sense of satisfaction of a job well done when someone approaches them with a problem, and they are able to resolve it. Others, generally those with strong technical backgrounds, would happily dispense with dealing with people at all. One indicated that he doesn't like the "Human Resources" aspect of his job. Another indicated that equipment doesn't call in sick, or quit

– as people have been known to do.

Almost one in three telecom managers cited dealing with invoice processing and verification as the one task that they would like to get rid of the most. The main source of frustration was not usually the processing, but the need to check the amounts being charged. To echo Marilyn from the call center case study, "Why can't a bill ever walk through the door and be right?" This is very fitting, considering that the variety of work is a major driver, yet processing and verifying invoices tends to be mundane, and often involves repetitive steps. Fortunately, there are software tools and outside services now available to streamline much of this administrative work, which are discussed in detail in Section III of this book.

John Holmen, a telecom manager with a Midwestern US-based healthcare agency with about 30,000 employees, is a case in point. As with many telecom managers, for him the worst part of the job is dealing with the invoices, checking them for accuracy and requesting corrections. He'd love to get rid of that part of his job, but recognizes that it has to be done.

Bill verification is an area of his work in which his department is not as good as he'd like – and has recently taken steps to deal with this, through hiring a billing specialist. He believes that being pro-active in reviewing bills is one of the most important steps a telecom manager can take.

One in six could not come up with a task that they would like to get rid of. This is consistent with all the opinions in this survey and others that come to the same conclusion that telecom managers are generally very happy in their jobs.

Career-boosting suggestion: Thinking differently about your work can help make it even more rewarding for you. It's called "reframing."

For example, let's say you work for a hospital with an urgent need for more staff in its trauma section. If your work reviewing phone bills results in $X savings in a given month, take a bit of time to calculate how many salaries (or parts of salaries) this provides for the organization. Then, think of the people you'll be helping.

Although many of the survey participants mentioned that billing issues and dealing with vendors was their least favorite part of the job, Lori Riccitelli, a billing analyst at a financial company, says that she enjoys dealing with vendors to retrieve credits. "I enjoy getting into the details and the challenge of getting the vendors to do what I want. My boss doesn't call me a bulldog for nothing." Instead of dreading these disputes, she is motivated by thinking of them as a challenge to overcome.

You can use this with your staff as well – it will help give them a new perspective on their work and possibly renewed enthusiasm.

Chapter 5
Miscellaneous Advice

*What can you learn from your peers
in other organizations?*

One of the questions asked in the Avema survey was, "What advice would you give to your peers in other organizations?" This chapter summarizes the responses to that question. The responses were varied, although several people keyed in on VoIP, seeing as how it is changing the telecom landscape.

Learn as Much as Possible About VoIP

Some telecom people think that VoIP is currently over-hyped and for now, it means more trouble than benefits. Many are wary of the level of reliability in the data network.

The more forward-looking telecom managers can see a lot more VoIP in the future, and a lot less traditional telephony. If you have a couple of more decades to go in your career, or even more than a few years, you should absolutely be learning as much about VoIP as possible. Section IV of this book goes into more details about how this technology will change telecom management.

Here are some other comments from your peers about VoIP:

"Analyze the technology, and approach it as a business analyst."

"VoIP will bring a relationship with your data peers to a new level. There is a lot more involvement than ever before."

"Have an open mind on the migration to VoIP."

"VoIP is coming, so voice people have to get training, understand it, and keep learning."

Other Miscellaneous Advice From Your Peers

"Buy the most reliable equipment available, as it means less time diverted to fixing breakdowns."

"Don't get discouraged, keep it simple, know your technology."

"Keep your systems and records updated, and keep up with training."

"Standardize, consolidate, take control."

"Focus on stability, contribute on success of company, effective customer service to internal client, keep thinking."

"Be good at customer service, continue to train and explain how telecom is needed and how to manage telecom and data.

"Be cautious of vendors, be careful of the 'too good to be true' statement."

"Proceed with caution on convergence."

"Do your homework, do it again, make sure you have all

the facts. Don't settle for the cheapest."

"Network, talk to people, communicate with the data group."

"Take emotion out of your job."

"Embrace change, let your company notice you & realize your value, have confidence."

"Find ways to show value, and measure efficiency."

"Document well for installation, changes, and reconcile circuits with billing."

"Do not jump at new technology, slow down, watch how it rolls out."

"Make your management aware of your expertise and educate them so they stayed informed."

"Don't let stress kill, be open minded."

"Keep up with technologies and the industry."

"Give power to people to do the work."

"Leverage your vendors as much as you can to do day-to-day support."

Career Management Thoughts From a Veteran

As a frequent participant in telecom-related e-mail discussion groups, I love the generosity with which some of the members of the industry will dispense wisdom on technical and other matters. This post came from Ken Steinhoff, Telecommunications Manager at Palm Beach Newspapers, Inc.

Ken was involved in a discussion about how telecom managers can demonstrate their value to the rest of the organization Consider:

It depends on how you define yourself. If you think of yourself as strictly a dialtone guy, then that's all you'll be and that's how your worth to the company will be perceived.

We have a rule in our telecom department that we don't give people what they ask for, we give them what they need. That means taking the extra time to ask what they are trying to accomplish. We may be able to give them something with more functionality at less cost that what they have asked for.

It's a cop-out to say that telecom doesn't have to worry about costs because the expenses go to an end-user department. End users are busy stamping out widgets or treating sick people or getting news in the paper. They have neither the inclination nor the background to make sense of phone reports and bills.

As I've mentioned in this forum before, we run daily, weekly, monthly and bi-monthly reports that help spot problems that aren't necessarily directly dialtone-related. An employee who spends, literally, 60 hours a month on the phone to a handful of local personal numbers isn't running up huge phone bills, but they certainly aren't productive employees. We flag patterns like those for managers.

*Conversely, as recently as last week, we had a manger call who thought that an employee was "spending too much time on the phone." Turned out that the person's average talk time was 30 seconds less than average. In the same week, another manager thought one of her employees was abusing the company's inbound toll-free line. An analysis of that department showed that, indeed, the person **did** answer almost 50 percent more calls, but there was no pattern that they were coming from the same number, nor were* *they abnormally long. She was just more conscientious about answering the phone than the other folks in the department. That's the kind of result I like.*

The folks in our department take the global view, whether it's helping users determine the most effective use of their physical space (we maintain the AutoCAD floorplans), cost containment, dialtone or infrastructure. We've come a long way from my first days in telecom when I asked for another body and was asked, "What do you need, just another gorilla to pull wire?"

Section II

Tactics for Delivering ROI

Chapter 6

What Can Happen
If Costs Are Not Managed

*What would happen if your organization
spends more than it should on telecom?*

As important as managing telecom costs is in a large corporation, it may be under even more scrutiny in government organizations. Major Canadian newspapers ran stories in March 2005 with headlines like "Six-figure waste on city phones" about the City of Toronto's telecom costs. The country's largest-circulation newspaper, *The Toronto Star*, reported: "In a damning report made public yesterday, auditor general Jeffrey Griffiths revealed that no one has been keeping proper track of rates being charged for telephone, cellphone and Internet service."

Considering that Toronto has a total budget of C$20 million (about US$17 million) per year in telecom, the amounts cited were not actually that big, and actually quite typical of large organizations. The types of billing errors, such as being charged higher long distance rates than the carrier contract, are common in any large organization. However, since government officials and taxpayers are not aware of how much work the telecom department does to keep the communications systems running, emergency services, and other mission critical work, the telecom managers were publicly blasted for being careless. The telecom department became a scapegoat for government overspending.

Here's how this story was reported:

> *Holyday* (a councilor) *thinks city council should demand that heads roll.*
>
> *"This kind of sloppiness just can't be tolerated," he said. "I think whoever's responsible for this should pay the price. It's just not good enough to keep throwing good money down the drain all the time."*
>
> *He said he's not looking forward to explaining the report to voters. "We have to answer to the shareholders and in this case, that's the residents and the taxpayers, and they're not going to be happy that this money has been wasted."*

In many organizations, the oversight of telecom costs is not managed by any single department. The telecom department plans and purchases services, contracts are handled through the legal department, and invoices are paid by the accounting department. There is often very little co-ordination between these departments, and often, there is no job description that states "look after the phone costs." In the past few years, larger organizations have become more aware of the need to manage their telecom costs, and there are more individuals dedicated to focusing on these areas.

The City of Toronto has telecom services in 900 locations, that get billed on 2900 different accounts, for 27,000 landline telephones and 11,000 wireless devices. Until recently, specialized tools to manage this massive amount of invoices and data have not been readily available. The Auditor General's report mentions that the city actually built some of its own inventory systems, but they were not capable of handling the vast amount of details, and were not designed to reconcile against invoices.

The recommendations made in the Auditor General's report were a very good start to gaining control and reducing the telecom expenses. Among them, here are some proactive steps that are being taken:

- Contract management is key to managing costs. This involves negotiating the most optimal pricing structure and terms, and then ensuring that invoiced charges match contract pricing.
- Ensure that cellular billing information is provided to all departments. Departments are to ensure that controls are in place for recovery of non-business calls.
- Review the use of cell phones by all City departments, and deactivate unused ones.
- Implement policies for acquisition of cellular phones, telephone, long distance, etc.
- Send out inventory information for landlines, data circuits, and cellular phones to every department on an annual basis, and have them verify that all the services are actually in use.

What else could be done to reduce the city's costs? The next few chapters summarize Avema's 10 years of experience in helping large organizations reduce their telecom expenses.

If you're an experienced telecom manager, much of this may seem familiar. You can use this next section as a checklist of pointers that may help you improve your ROI.

Chapter 7

Reduce Vendor Charges

Even though you may have already reduced costs significantly, is there still room for improvement?

Gather Information And Do Your Homework

Consolidate Vendors and Invoices

The fewer vendors that you have, the easier it is to manage them, not to mention billing! Also, if you can commit to more volume with fewer vendors, you can get better pricing. The first thing to do when your organization merges or acquires another is to reduce the number of vendors as much as possible.

Also, reducing the number of billing accounts will help in processing and auditing your invoices. It also makes it easier to calculate total volumes to give you more leverage in negotiations.

If your organization has multiple divisions or sites, make sure that all the telecommunications invoices come to the same place, and that contracts are all handled centrally. It may still be important that various parts of the company receive data on their telecom usage, but having all the information come in to one place can have a large effect on cost management. You can still get the telecom usage information out to your various departments afterwards.

Electronic Billing

Another useful method of gathering data to analyze spending and usage patterns is to get your billing electronically whenever possible. Having fewer vendors will also help here, as each one has their own unique billing format. Some vendors have multiple electronic billing formats, particularly those that have been amalgamated from numerous smaller companies over the years. Typically, vendors provide their own proprietary software to analyze the billing data. Minimizing the number of electronic billing formats that you need to work with is another good reason for consolidation wherever possible.

Some carriers are charging customers for electronic billing, which may seem counterintuitive. Carriers also reduce their costs by providing electronic billing instead of paper. You can use this to your advantage to negotiate. If you spend more than even a few hundred dollars per month with any carrier, you should be able to get them to waive the fee for electronic bills.

Unfortunately, some smaller regional vendors may not be able to supply billing in electronic formats, so you may still be stuck with some paper.

Negotiate the Best Contracts

Some managers enjoy vendor negotiations, such as John at a Midwestern health agency, "I love the art of the deal," he says. He enjoys working with vendors to help his organization get the value it wants, structured in a way that gives the vendors the business they want. With $5 million to $10 million in new contracts to negotiate each year, this gives him plenty of scope to use his negotiation skills. "That's the challenging part and the fun part," he says.

Others said that they find "beating up vendors on price" –

to use an expression from one survey respondent – to be fun in a combative sort of way.

I find it very progressive that John understands that there needs to be a win-win in these negotiations. He knows it's important to help vendors get what they want – more traffic, longer terms, a positive referral, or a promise of future business.

Commodity Services Mean Lower Prices

Many telecom services have become a commodity. There is virtually no differentiation from one long distance service to another, or even Internet access from multiple providers with the same technical specifications. This has caused pricing to come down dramatically over the years, and it continues to come down. This is the lowest hanging fruit available to increasing your telecom management ROI.

The key to obtaining the best possible pricing and terms is to have as much leverage on your side as possible. Understanding your spend and usage patterns, and which vendors are best suited to those patterns will go a long way to reducing your costs.

If you have a purchasing department, they can help with the process of negotiations, and will generally help you get better pricing. Couple the purchasing expertise with your own domain expertise, and that of your peers in other organizations. Find out what the current competitive rates are so that you can benchmark vendor proposals against the best of breed.

You might also consider enlisting help from experts who specialize in negotiating telecom contracts, since they will always be up to date on the latest market rates and the terms that are available and realistic for your organization.

Simple Price Structure Is Better

Whenever possible, negotiate to get the simplest possible pricing structure. As an example, in Canada, flat rates for long distance across North America have been the norm for many years. It costs the same per minute rate to call from Toronto to a neighboring suburb, to Los Angeles, or to Alaska. This happened because of intense competition, and insistence from businesses and consumers to make pricing simpler.

For Americans, however, you are likely still stuck with regulation that dictates certain tariffs for different types of calls. Working on getting the simplest possible price structure will help you validate those invoices, and help you manage your costs in the long term. It may even be worth paying a small premium.

Simple Pricing For Wireless Is Even More Important

With volume wireless contracts, there are currently several pricing schemes available including individual price plans for each user, pooled pricing, and flat rate pricing.

Individual price plans mean that in order to maximize your savings, you will need to choose the lowest cost plan for each user, depending on their habitual usage. It also means that you have to keep an eye on changing usage patterns and continuously adjusting these plans, which can be a lot of administration. There is software available to help in this area, however, it can still be a lot of work.

Pooled plans mean that you can have several users all sharing a "pool" of minutes. This is beneficial because you wouldn't need to worry about constantly monitoring and changing users' plans. Even with this scenario, it is still important to monitor your overall usage and adjust the pool as necessary.

Flat rate pricing is more common outside of the USA, where a monthly fee is charged for each user, and then a flat rate per minute is charged for usage. This makes monitoring usage much simpler, and again, there is no need to monitor and change plans.

There is not necessarily a "best solution" out of these options for everyone. The optimum pricing structure really depends on your organization's usage patterns. For example, trying to optimize price plans for each individual user may be a losing proposition if your users' spending patterns fluctuate dramatically from month to month because you will always be a step behind. On the other hand, if some users in your organization regularly use large volumes of off-peak minutes, setting them up with individual plans that include unlimited off-peak minutes may save you enormous amounts. Roaming and long distance are also items to consider when you analyze which pricing structure works best for you. Pooled and flat rate pricing tend to be priced a bit higher, but it may be worth paying if it saves you extra costs in managing plans. Also, it may cost less anyway, depending on your usage patterns.

Another option is to use one pricing structure for some of your users, and another pricing structure, or even another vendor altogether, for another group of users that have different calling patterns. For example, you may have the majority of your users in a pool if the numbers work out better, but for some heavy-duty users or people who travel extensively and incur roaming costs, you might optimize those users' plans individually.

Ultimately, you still need to do your homework to determine which pricing structure works best for your organization by analyzing usage patterns and volumes.

Other Pricing Considerations for Wireless

With volume contracts, you can also sometimes negotiate having monthly features such as voice mail or caller ID included for every unit. Even though the cost savings is only a few dollars per phone, it can add up significantly on a large number of units.

Mobile-to-mobile calls within the same wireless network, or within your corporate account are often included in corporate pricing. If your users are calling each other frequently, this may save you money. The push-to-talk or direct connect services offered by some carriers typically offer some sort of flat-rate pricing for frequent users.

You can also sometimes negotiate "free" calls to one or more pre-defined numbers. If your users are constantly calling back to the head office, this can save you a lot.

Be sure to include pricing for future hardware purchases in your negotiations. You wouldn't want to get stuck paying full retail price for a new mobile phone when one of your users loses one. It can be aggravating to know that any consumer could purchase the same phone for a fraction as you, despite your large volume.

How Close Are You to Your Vendors?

Having tight relationships with your vendors is a double-edged sword. On one hand, it may make negotiations more pleasant, and you may get better service from them. On the other hand, you may have far less leverage when negotiating. I've seen pricing for several companies that "partnered" with a major telco for marketing and other non-telecom related aspects, and they were paying between 50% to 200% more than the competitive market rates at the time. This was costing them a six-figure annual amount more than necessary. They weren't obligated in any way to pay more than anyone else, but the telco had taken advantage of the situation.

If you do have preferred vendors, you can still keep them honest by issuing RFPs and doing proper negotiations. Knowing that you have the opportunity to switch your services to another competitor is often enough to get them to work with you on getting competitive pricing.

Other Terms Besides Getting the Lowest Rates

Sometimes there are even more important terms that need to be in your contract than just getting the lowest rates. For example, if you're signing a contract with a term of several years, you want to make sure that you have the ability to continue to obtain best market pricing throughout the duration of the contract. One Internet service provider told me how they signed a 5-year contract with their incumbent telco for data services that were the core of their Internet offering. This had been costing several thousands of dollars every month, but 3 years into the contract, alternate technology was available that only cost $69 per month. Because they did not have that one clause in their contract, they were stuck paying close to $100,000 more than necessary over the term of the agreement.

You may be taking over a situation like this with terms that are not to your liking. If you find yourself in a position like that, you may be able to negotiate with your vendors anyway. Carriers are always concerned about "churn," that is the rate at which customers leave them for the competition. If your company is stuck with the prospect of paying $100,000 more than necessary, it doesn't hurt to ask if you can go around the contract. If you've only got one year left in the agreement, use it as leverage to re-negotiate a newer contract with a longer term than the remaining one year, but at lower rates. Let the carrier know that if they stick to the letter of the previous contract, charging you much more than the competitive market, that you will definitely not be signing with them when it comes time to source a new contract.

Don't Cut It Too Close

It is worthwhile to note that pushing vendors to reduce their prices can eventually become non-productive. If you've been reading the telecom press, or even the regular business news, then you know that many major telcos are in financial difficulty. For years, they have been cutting back on operations costs, including customer service. Some of them track the profitability of each major account, and will make more resources available to those where they're making more money.

In any case, prices can only come down so much, and so you can only rely on this tactic to get you a certain amount of ROI. The next time your executives ask you to cut costs, you need to have a plan ready to address how you will go beyond price cuts.

Audit Your Bills

Even if you're not a telecom manager, you undoubtedly know that telcos make many billing errors. Technology analysts such as Gartner, Meta Group, and Aberdeen estimate that billing errors account for anywhere from 5 to 12% of all telecom charges. For a company that spends one hundred million per year, that's an enormous amount of money!

How is it that telcos manage to mess up their billing to this degree? There are a number of factors. The software that is used in most major telcos dates back several decades. As new billing requirements such as changes in taxes and fees, new pricing structures, and new service offerings evolved over the years, the software didn't necessarily keep up. Also, the billing software still must rely on manual processes, which are prone to human error. For example, when you place an order to cancel a service, it may be discontinued, but it doesn't necessarily mean that it will be removed from the bill. Telecom managers must be vigilant of some of these areas that are most likely to result in billing errors.

One error that I have witnessed repeatedly is that a new contract is signed with newer low rates to take effect on a certain date. Since it may take some time for these new rates to be updated in the billing system, it could be several weeks or months before you actually benefit from the reduction in rates. If the change in your pricing is significant enough, make sure that the carrier backdates those rates to when the change was supposed to take place.

Even more common in multi-site organizations, if you have multiple billing numbers, you may not be receiving the correct rates across all of your accounts. Since many of the telco billing systems require each account to be updated individually, it is easy for some to slip through the cracks. This happens frequently to companies that have a complex structure with many corporate entities, names, and divisions that are all supposed to benefit from your central contracts. Some of these are harder to catch, since it's most commonly the smaller offices with the lowest amounts of spend that are missed, but when they're all added up, it can be a significant amount of overcharges.

One client that we at Avema performed a cost review study

for had moved to a brand new office two years prior to our audit. Even though it was only a relatively small company of a few hundred people, the telco had forgotten to apply their long distance rates to the new accounts. Instead, this company was charged the default non-discounted rates that were being charged 15 years ago, in the neighborhood of 30 to 40 cents per minute for every call in North America. We retrieved a refund of almost $700,000 for this one billing error.

When To Use Outside Help

If you've been in telecom management for any significant length of time, you will have been contacted numerous times by bill auditing companies, offering to help you save money by negotiating costs and auditing your invoices for billing errors in order to get refunds.

Having started out as a consulting firm, Avema has helped many companies with this type of work. Most of our clients have been large companies, with sophisticated telecom departments. Because of our specialization in cost reduction and recovery, we were able to help our clients maximize their results.

Now, outside help is not always useful or recommended. I found that the most successful telecom departments used our services where they needed extra resources, for instance, if they were tied up with other projects, but still needed to do a cost reduction exercise, or if they had already gotten 80% of the results that took 20% of the time to do. They were able to use our expertise and tools to get the remaining 20% of the results, and we were able to do it in an efficient manner because of our specialization.

Sometimes, it's helpful just to get a second set of eyes, a fresh point of view. For example, we've had many clients do as much as they can with contract negotiations, and then have us review what the vendors were offering. Sometimes, we were able to help them with some fine points that ended up saving them large amounts.

Chapter 8

Reduce the Amount of Services

Can you think of any services that your organization doesn't need to be paying for?

Beyond getting your pricing and contracts in order, the only other way to reduce telecom costs is to reduce the amount of services that you purchase.

Of course, this doesn't mean that you should go and cancel all your telephone lines. While that would save you money, it probably wouldn't be a smart career move. But there are other ways to reduce the amount of services that you use.

Infrastructure Reduction

In at least half of the WAN networks that Avema has analyzed, there have been significant cost savings because of excessive circuits or bandwidth. It's a good idea to review the network design every once in a while, depending on how frequently changes are made. Otherwise, you may have a number of inefficiencies in the way your network is configured.

I've been surprised by how often we review a client's infrastructure and billing, and find out that they're still paying for circuits to a building where they moved away from two years ago, but never placed a cancellation order. In a case like this, a credit would be much harder to obtain, since it's not the telco's fault.

Over the years, as your organization grows, restructures, and maybe even shrinks occasionally, phone lines and circuits get left behind and forgotten. If you're in a very large organization, it can be difficult to find the time to perform a periodic audit to find services that are still being paid for, but not used. This can, however, be a very worthwhile exercise.

Traditionally, most companies have handled these tasks in a reactive way, performing an audit of invoices and services periodically to find charges for circuits that were cancelled a long time ago, or services that are no longer used that you don't need to pay for. Today, the trend is moving towards being more proactive, which can save you large amounts of time tracking down billing and haggling with vendors over an event that happened two years ago.

Keep a Clean Inventory

Maintaining a list of your various circuits and associated services, an inventory, can go a long way towards managing these costs proactively. Many companies already do this in one way or another, most typically just keeping a list in a spreadsheet. If you don't already have an inventory, there may be a fair amount of work up front, getting the list together in the first place. Just having an inventory, however, does not guarantee that your services will become straightforward to manage. The single biggest challenge is in keeping that inventory up to date. In a typically frenzied telecom environment, orders are being placed all the time, and it is easy for anybody to make the mistake of forgetting to update the inventory. The inventory can easily become out of date, making it useless. Even if it is updated, there still needs to be a way to reconcile your incoming invoices against it. Section III of this book explains in more depth how you can use software tools to help with this ordeal.

Reduce Your Organization's Usage

Now that you've got your fixed monthly costs as low as can be, there are still opportunities to reduce usage of services like long distance and wireless, but it may be more difficult for you to control directly, since it's other people who are actually using these services.

Since this type of cost control is more about organizational behavior, it is important that you have full support from senior management. If you haven't already implemented these types of cost controls, it is a great way to get more visibility and to demonstrate your proactive approach to delivering ROI in your organization.

Having telecom costs, at least usage-based ones, charged back to the department that incurred them can create accountability and responsibility for those costs. Otherwise, department users never think of what the telecom costs are, and certainly aren't conscious about reducing unnecessary usage or expense.

The telecom manager at a hospital says that one of the cost-reduction steps her organization takes is to have even straightforward MACs require approval at the Vice-President level, so that every expense gets reviewed in order to verify if it's really necessary. She also helps her departments with educational efforts so they can see where their telecom costs are, and where they may be able to cut.

Showing each department how much they use for telecom services is a common practice, although there are several ways that companies perform this function. The simplest method is to distribute a report showing the total dollar amounts charged back, however, this doesn't contain enough information to allow your end users to make any intelligent analysis or decisions based on it, other than keeping an eye to see if that number increase or decreases. Also, it tends to create more work for you, since they will inevitably call and ask for more details.

One telecom manager showed me a printout that must have been about 100 pages long, detailing telecom spend across different departments. Considering that this company only spent about $2 million per year on telecom, this was a case of overkill. Apparently, senior management had decided that this was the best way to report telecom costs, and the telecom manager got stuck implementing it. This was sent out, on paper, to numerous department heads. Unfortunately, the telecom manager admitted that the system was not as accurate as it could be, and it was extremely time consuming. Also, because the report was so long and detailed, it hardly ever got read. This is an example of the right idea, but the wrong execution.

This is an area where having efficient processes and software in place to automate the function can not only save you a lot of time, but also provide more accurate and useful information to your users. Ideally, you want to give your users access to the most complete information, yet avoid overloading them with useless facts and numbers. Section III in this book describes in more detail how this can benefit your organization, such as enabling self-service reporting and analytics so that your users can find out exactly what they want to, at their convenience.

In terms of reporting and making your role more visible, it's a good idea to put your department's name, and maybe your own name, on each report that goes out. Even if you don't have time to directly speak with your end users every month, this helps to remind them that these reports don't appear out of thin air, that you put work into getting this information out to them.

Special Considerations for Wireless Usage Control

Controlling usage costs for wireless devices in particular can have a big effect. The difference with these devices is that people use them out of sight from the office and their managers. This creates the temptation to use them for personal reasons more than a landline phone, and it also creates the suspicion from senior executives that they are being abused (and sometimes, they're right!)

A large newspaper company's CIO described how the CEO's pet peeve was the wireless costs. Even though the CIO said that she had better and more important things to do, she had to spend time focusing on getting wireless costs under control, because "Tony" was unhappy about them.

Earlier on, cellular phones didn't necessarily belong to the telecom department, or any other. Since the costs have been getting so high in the last few years, more and more telecom departments are being given the responsibility to manage this. This is another area where there is a significant opportunity to demonstrate quick ROI, since there are numerous ways that wireless costs and usage can be reduced.

When cellular phones were first introduced, they were only used for the most urgent calls. As wireless devices became more common in everyday life and usage rates came down, most people became accustomed to using them regularly, sometimes much more than necessary. In business, the risk is that they are used largely for personal reasons, rather than as a productivity tool.

An extreme example: one company in the UK banned personal calls on mobile phones, and their costs dropped by 75%! We do not recommend taking measures that will be seen as excessively harsh, but this does demonstrate how much costs can get out of control if they're not managed. Often, just the knowledge that costs and usage are being monitored is enough to motivate people to be mindful. With landlines, call accounting has been used for many years in this way.

Even when used exclusively for business, employees may use wireless devices in the most convenient way, rather than the most cost effective way. For example, roaming overseas can be particularly expensive, and users must change their habits in order to avoid excessive charges.

With the more recent adoption and increasing use of mobile internet and data services, companies risk losing control of these costs as well, unless spend policies are put in place.

What Doesn't Work

Most organizations negotiate volume contracts with their vendors, and are billed centrally for wireless services. However, some companies have tried to get employees to handle their own bills, and expense the costs back to the company. This tactic that is intended to cut costs often ends up *increasing* costs, as much as 100%. One company with 500 mobile phones attempted to reduce costs by changing from centralized corporate paid invoices to employee expense reports, but instead saw their costs double within 6 months, from $500,000 per year to $1,000,000!

A major disadvantage is that the company no longer has visibility into how mobile services are being used, since they no longer collect the billing data. How would you know if the $200 limit that you assign is more than they actually need? More importantly, individual employees are not able to take advantage of corporate volume pricing or pooled plans, which can significantly affect the rates.

There is merit in the theory that employees will watch their own invoices more diligently, and limit the amount they spend. In practice though, employees have better things to do than review every detail of their wireless bill, trying to get the most cost effective pricing. It is beneficial to show them the costs and usage that they are responsible for, however, invoices should still be paid and processed centrally. With the ability to distribute portions of the invoice that are relevant to users, you get the best of both worlds.

Best Practices of Spend Policies

From discussions with dozens of large companies struggling with their wireless costs in various regions, Avema has gathered vast data on best practices of reining in wireless costs for both voice and data devices. Our software is designed specifically to address these challenges. This guide covers one of the more important aspects; implementing spend policies, monitoring, and enforcing them.

It is imperative to have executive level support, because your users must understand that they are accountable, and that the policies will be enforced.

1) Group users by function and expected usage

For instance, field service staff will have different spend patterns than salespeople or executives. Also, within each function, there may be subgroups, e.g. field staff in different territories may have different requirements for airtime, long distance, and roaming usage. Some groups may require data services, whether it be for their laptop or PDA. You may even find that some users who are assigned mobile devices rarely use them, if ever, and do not actually require them for their job function.

To gather information, you may want to discuss this with the various business units in your organization. You can do some preliminary work and backup analysis by reviewing historical billing information. It is best to look through several months of billing, as there may be anomalies in a sample of one or two months. Of course, having your billing in an electronic format helps tremendously with this task, and having the right software tools will make this much easier on you.

2) Compare individuals within groups

Within these groups, compare spend and usage between individuals to determine what the typical usage should be. Correlate this data with performance and effectiveness of your staff. For instance, a salesperson with a higher volume of calls may be very reasonable if he or she is bringing in twice as much revenue as any other salesperson. On the other hand, if that salesperson is generating half as much revenue, then he or she may be spending too much time chatting on the phone instead of selling!

3) Standardize ordering of hardware and features

New models of wireless devices are being created and marketed every day. Since these gadgets have become so ingrained in our culture, your users may want the latest and most expensive equipment available. But do they really need a phone with polyphonic ringtones, an MP3 player, or a camera? Determine standards for the models of phones or PDAs, and which accessories and features are required for your user groups' business functions. Take into consideration whether some users require both a phone and a PDA, or if there are devices that your company can standardize on that incorporate both.

Establish centralized ordering so that every user must request new equipment or features through the same process. With the right software and process, you can define exactly what can be ordered, and by which groups of users.

4) Spot check call details

Especially for high volume users, have a look through some of the call details to see if their calling patterns make sense in the context of their job function. Watch in particular for excessive personal call usage. Are there hour-long calls that could easily have been made at no cost or a lower cost on a landline? Do they call directory assistance 30 times in a month at $1.25 each time? Are roaming calls frequently made when traveling overseas, when a calling card or payphone would be significantly less expensive? Again, having electronic data in a useful format will help you tremendously here.

5) Determine reasonable thresholds for usage, and consequences

For each group, estimate what a reasonable amount of usage should be and set a limit. You will want to ensure that it's low enough to contain costs, but high enough so that your employees can perform their job functions without worrying too much about crossing that limit. Also, you can expect that there will be exceptions to manage, where an employee spends more than the limit, but for legitimate reasons.

Establish the consequences for exceeding these limits, and communicate this to your users. Consequences may include having the user justify the excessive usage with their manager, or personally reimbursing the company. It's important that the consequences have enough 'teeth' to make your users feel accountable.

6) Monitor usage

With the threshold limits defined, you will want to monitor for any usage that exceeds them. This may entail simply running through the electronic billing data, keeping track in a spreadsheet, or using sophisticated software that can automatically notify you of any discrepancies via email. With the right software, you can also watch for other potential issues, such as calls to specific countries or numbers, calls over a certain duration or cost, or charges for downloading ringtones and games. When these discrepancies arise, define how they should be handled. Do you automatically forward an email to the employee, and copy his or her manager, or the controller? What is the tone of this message?

7) Send regular reports

Communicate the usage and cost information to individual users. Some companies have experienced immediate cost reductions of 30% or more, simply by making their users more aware of their usage and associated costs.

If you have paper billing, you can distribute the portion of the invoice that they are responsible for. A much more efficient way to distribute this information is to use a web-based system that allows employees to log on and view their details. A monthly email can go out with a summary of their usage, reminding them that the detail of the bills is available online. Letting your users view and manage their own billing information can be very effective. This must be balanced by providing them with the software tools to perform these tasks efficiently.

As well as advising them of their wireless charges, it is also important to give them information on how they can reduce their charges. Focus on the costs where there is a significant ability to make a difference. Some people use Directory Assistance reflexively without thinking of how these charges add up, when they could easily use a phone book or an on-line White Pages directory. Or, they may use their mobile phones while away on business, racking up serious roaming charges, without thinking that using the room phone or pay phone, plus a calling card, would be much lower in cost.

Optional Step: Tracking Personal Calls

One more step that some organizations use is the reimbursement of personal calls. This involves tracking any personal calls that are made to certain phone numbers, such as a spouse's office number, or child's school. Before sophisticated software was available, this would have been done on paper, an onerous task for any employee to page through their phone bill line by line. As you can imagine, this could backfire and create a disgruntled work force.

With the right software, however, this can be mostly automated. Each user can be given individual access to his or her billing information, summarized by the telephone number called. These numbers can be marked once, and remembered for future bills. As long as this list of phone numbers is kept accurate, the system can tabulate the costs, and produce a summary of how much each user should pay.

Case Study in Applying Wireless Spend Policies

These steps were used to implement spend policies at a construction company, where mobile costs were rampant because every employee was assigned a wireless phone. Worse, many of their employees relied on that phone as their primary means of personal communications.

An Avema Global Alliance partner, CPS, conducted a review that included 18 months of invoices to determine the spend patterns, and make recommendations for usage limits. Together with the client company, CPS grouped employees by their job function, and a number of airtime minutes was assigned to each group. Also, they discovered that there were significant costs associated with directory assistance, text messaging, and some roaming, even though there were no legitimate business reasons for them.

The client decided that the consequences of exceeding these policies would be to have the employee's paycheck debited. CPS created a form for each employee to sign, acknowledging their acceptance of the policy and its consequences. Every month, the bills are monitored for usage beyond the threshold limits as well as the extra miscellaneous charges. These exceptions are then forwarded to the accounting department, which enforces the policies by deducting the excess charges from the employees' paychecks, and including a statement showing their usage.

Before the policies were implemented, CPS had optimized their pricing by configuring price plans and negotiating with their carriers. Even after this had been done, wireless costs were reduced by a further 27% through these policies. Also, employee productivity may have increased, as some individuals are now using half as many minutes as before.

Chapter 9

Revenue and Productivity Enhancement

What is your organization's strategic direction, and how does your work as a telecom manager support this?

Now that we've covered cost savings, what about helping your organization to drive more revenue, i.e. sales? Most telecom managers do not think of their job as helping to enhance revenue, and only think of themselves as a behind-the-scenes support function. In many cases, it is true that telecom managers do not affect revenues, but for those who do, their skills can be considered much more valuable.

Admittedly, it is more difficult to come up with revenue enhancing ideas, and this will depend largely on your industry. Most of your efforts in this area are more likely to have an indirect effect, rather than directly increasing sales the way the sales and marketing departments do. Your role then, is to help the top producers in your organization to be more effective in their jobs. Here are some examples.

The Well-Equipped Salesforce

In most organizations, salespeople are the most obvious revenue generators. Executives at virtually every company on the planet likes to hear about increased sales. Although you wouldn't necessarily go out to sales meetings with your company's representatives, there are other ways that you can help the sales effort.

How valuable is it for a sales manager to be able to compare the revenue generated from the top salesperson with the average or lowest performers, and see how the volume of calls, length of calls, etc. compares? Outside of a call center, this can be hard for a sales manager to access. In fact, I would bet that most sales managers aren't even aware that they can get useful information from call details. Either helping them to do the number crunching, or providing them with intuitive software tools with which they can perform some analysis themselves can make a big difference.

How about working with the IT department to help provide the road warriors with an office in a laptop or PDA, connected in real-time with the main office? This may include having live access to real-time inventory records so that when a salesperson sells 1,000 widgets they can tell the customer if the product will be delivered next day or next month, depending on whether it is in stock now or will need to be specially manufactured. Salespeople should also be able to interact with the production people so they can assure the customer, "This order is scheduled to be produced next Thursday afternoon, so we can ship it by overnight delivery and you'll have it Friday morning."

If telecom managers work closely with executives responsible for bringing in revenue, all of a sudden your job becomes a very important function, which could actually have an influence on the growth of the company.

Efficient Telecommuters

Telecommuting is a growing trend in many companies, as it offers several benefits to both employer and employees. For employees, having the flexibility of working from home, at least some days each week, and cutting down on their commute may provide more job satisfaction and increase their productivity. For employers, it might mean reducing the amount of office space required.

The challenge is in making sure that these home-based workers have the telecommunications tools they need, and being able to manage these services and costs effectively. This might include internet access, a separate phone line for their home office which is eligible for the same low long-distance rates as the calls that originate off the organization's PBX, or even an extension of the PBX through VoIP, with all the office phone features. Calls to their corporate office number need to be routed seamlessly to their home office.

Frustrating Headhunters

Another problem might involve employee retention – too many good staff are being poached away by other organizations, possibly competitors. Many of these employment offers come in via phone calls from recruiters. By developing a way to flag calls from established recruiters' offices, these calls can either be blocked, or at least trigger some investigation if the volume of calls increases. The result is that recruiters have a more difficult time contacting employees and making them offers, and managers can be more aware of employment trends.

While this will not solve the problem entirely, it can go a long way towards helping solve a serious problem for the organization – and will get the telecom manager noticed as an imaginative person who is a good source of solutions.

The Flexible Organization

Many employees today work in flexible teams that are brought together for a project, with the team being disbanded when the work is finished. Employees then join other teams. It has been found that geographic proximity is a big part of success for this type of arrangement.

If you can make it easy for employees to move into the same part of the office for several weeks or months, and then move elsewhere as their new team forms, you will be contributing to this ideal. This means streamlining MACs as much as possible, so that employees can move to a different office and immediately use the same extension or DID number as before. It may be worthwhile to assess whether your organization would benefit from IP telephony to streamline this. If your company has employees that move around frequently, you will be seen as supporting an important corporate goal.

Strike-Proof Company

A pizza chain, which relies heavily on phoned-in orders from customers, shows how telecom can help meet strategic goals. This company experienced a strike in its call center, essentially shutting down all sales.

Following that painful episode, the pizza company decided to find ways to become less vulnerable to labor unrest. By having its order-takers work from their homes, they would be less accessible to labor organizers. Also, having people work from home has given the company access to a labor pool of people who prefer to work part-time and from home, and this has helped in cutting costs.

This flexible arrangement can help home-based order-takers to be as efficient as those based in a call center would be.

Nurses On Call

Some hospitals are beginning to install wireless devices in their buildings, with data on the go, and voice over a wireless LAN. Since doctors and nurses are constantly moving around inside the building, this can have a tremendous effect on their productivity. Instead of having a nurse sent to a patient's room every time the call button is pressed, she or he can answer a wireless device to find out if it's an urgent situation, or if it's a routine chore that can wait until later. Doctors and nurses can more easily communicate with each other, saving time that can be used instead to provide better patient care.

Help Other Business Units
With Relevant Information

In addition to helping the company control its usage and costs, usage reporting and analysis can have much more value as a management tool, which ultimately results in higher productivity and revenues. For example, excessive personal calls made by an employee while on company time may be an indicator of a poor work ethic, which may point to larger issues. Considering that this person is being paid for their time and effort, using that time for personal reasons can even be considered a form of theft. This can trigger a less-than-favorable performance review and can provide unassailable data to protect the employer in the case of wrongful-dismissal suits.

An employee who abuses phone privileges may be doing it because she or he is bored or in the wrong job – valuable information at appraisal time, when managers can use this information to probe to find out about what would make the employee more interested in achieving higher performance.

Ken Steinhoff says that he's been with his current employer long enough to spot anomalies and problems. For example, he might see that one employee seems to be making a large number of calls to a specific South American country code, such as that of Colombia. He might investigate and finds out that this is entirely appropriate – this is a reporter who specializes in the Latin American beat.

Then he discovers that another phone is also being used to make many lengthy calls to Colombia. Only, these calls are mostly to one number, and are being made well after usual business hours, and happen to be from a cubicle in a secluded part of the office. He investigates, and finds that a staff member with family in Colombia has been keeping in touch with home, at the company's expense.

He says that such investigations as this have given him a reputation around the office for watching closely to see if there are unauthorized phone expenditures. He doesn't actually monitor every bill, he says, just enough to give the impression among the staff that he is, and that unauthorized calls are just too risky.

Sometimes a manager will come to him with the request to monitor a specific individual's phone. He will decline and instead, suggest doing a check on the entire department. This way, one individual doesn't get singled out, and the department-wide check helps put each person's calling patterns into perspective.

Of course, employment decisions shouldn't be made solely on telephone usage records, but it is an important indicator that can help managers make better decisions.

Business Call Analysis

Legitimate business calls, likewise, can provide a wealth of management information. Use your software to generate reports on outbound traffic by factors such as customer or client, geographic area, time of day and other factors.

These reports might indicate that a high proportion of calls, and longer calls, might be to a certain customer – indicating that perhaps this customer is costing the organization more, in terms of employee time, than it is worth. The organization might consider raising prices for that customer or instituting a customer-service charge program.

In other cases, outbound call analysis might indicate that many calls are being made to a particular area code – possibly indicating that there is a great demand in that location for the organization's product or service, and that a branch office there might be worth establishing.

Data on inbound toll-free numbers are particularly useful. If promotional materials such as brochures for different product or service carry individual toll-free numbers, it is easy to generate reports indicating which published materials generated more inbound calls. This helps the marketing and advertising departments to fine tune their advertisements, and measure response, which can in turn help them to create more sales opportunities.

Since your end users in various departments may not be aware of the rich information that you can provide, find ways to initiate discussions and discover where you can help out.

Offer up some suggestions of what useful reports you can provide, and get some ideas on what they are most interested in.

Section III

Industrial Grade Power Tools

Chapter 10

Can Telecom Management Be Automated?

If you could automate some administrative tasks and save yourself many hours every week, what more useful work could you do with the extra time?

For the longest time, for telecom managers to keep accurate records of costs and check billing for errors, there was a lot of manual work involved. Because of all the many, many bills, and the billing errors, cost allocations, and everything else, telecom management meant a lot of administrative tasks.

Eventually, virtually all telecom management functions will involve the use of software tools to streamline processes. Instead of having heaps of phone bills on different desks, all invoice and inventory data is stored in one database. Information is easy to retrieve, invoices almost process themselves. IP phone systems may reduce the complexity of Moves/Adds/Changes. Can the entire department be automated?

The answer is no. And yes. Many mundane and repetitive chores can be automated, but there will always be a need for higher level thinking, strategic work, and using the technology to deliver the maximum results. Consider other functions in a typical company, like finance, where mammoth enterprise software has been installed in every large company. Today, there are far fewer clerical staff, and data is more accurate and easier to access than ever before. However, there are still numerous finance professionals who take a high-level view, and are able to make the most use of technology.

In the manufacturing sector, robotics and machines have replaced countless factory jobs that blue collar families used to depend on. Despite this, factories still employ many workers, engineers, supervisors, and sometimes just employees dedicated to monitoring the technology. Factory workers today are more likely to be highly paid specialists that design and fine-tune processes.

The same thing is now happening for telecom departments. There will certainly be fewer employees required, particularly clerical staff. However, like any other area of the company that has been changed by new advancements, there will always be a need for more strategic personnel, and those who install, run, and maintain the technology.

This newer technology, a suite of software, is like having heavy-duty power tools. Until recently, only hammers and screwdrivers have existed, and all this detailed work had to be done the hard way.

How much time does your department spend:

- Processing or verifying phone bills?
- Searching for, or compiling various bits of information that you need just to do your job?
- Creating reports and sending them to various other staff in your organization?

Most of this work is not exactly glamorous, and certainly

not fun. More importantly, this manual drudgery does not add any value to your organization.

Imagine: If you could automate these administrative tasks, how much more effective could you be at your job? Could you save the company money by being able to take on more, without hiring new staff? If you automate the work performed by clerical staff, could they be re-deployed somewhere else in your organization where they could add more value?

You may have already worked hard at getting the right processes and people in place, and now with effective technology that has only recently become available you can accomplish so much more. The right telecom cost management software can automate much of the invoice mess, notify you of any billing discrepancies, help you keep track of your orders and inventory, and enable your end users to pull up their own reports.

Low level administrative work can be reduced to managing by exception, only requiring your time and attention when a potential problem or opportunity arises. This frees up your time to deal with bigger-picture issues and add more value to the organization.

What Is Telecom Management Software?

Because of the many types of software applications available to help telecom departments run more efficiently, it is important to define the category of "telecom management software." For the purposes of this book, telecom management software refers to any specialized software tools that either provide analytical information, or streamline processes within the telecom department. These include applications such as call accounting, invoice management, inventory management, and others. Excluded are spreadsheets that are used to manage tele-com, and carrier-specific bill analysis software.

The earliest telecom management software was call account-ing in the late 1970s. Before PCs were widely available, this ran on mainframes.

Over the years, other specialized tools have become avail-able. In the 1990s, some inventory tracking software started to appear, and in the late 1990s and early 2000s, invoice man-agement, wireless management, and more integrated offerings have appeared.

Invoice management software was spawned from the tele-com auditing industry. In the USA alone, there are hundreds of firms that provide audits of telecom billing. Since there are no large investments to start up a telecom audit firm, many one and two person shops have sprung up in recent years. There are even companies that train would-be entrepreneurs to audit bills, and at least a couple of companies have tried to fran-chise the concept.

Some of the more successful audit firms evolved into out-sourced service providers, managing billing day to day for their customers, instead of snapshot audits every year or two. Meanwhile, other firms concentrated on developing software to automate billing management processes within large com-panies.

Some evolved from call accounting providers, as long dis-tance costs plummeted, and fewer companies were willing to pay for software that managed costs in the neighborhood of $0.03 per minute. Call accounting firms either acquired or partnered with invoice management software companies, or attempted to write their own software.

As the technology analysts covered this emerging market, numerous press releases were issued, and still more new software companies came out of the woodwork. As of the writing of this book, there are dozens of companies claiming to provide telecom expense management software and/or services.

Already, there has been some consolidation in this space, where deep-pocketed banks and venture capital firms have financed mergers and acquisitions. As for the software, there will continue to be improvements made by the most innovative development shops, and telecom management software will eventually consolidate into fewer applications that are more powerful than the individual components of the past. For example, there are currently several software firms that have developed applications to manage only wireless costs, and others that only manage inventory. In fact, it's possible to buy wireless management software from one provider, call accounting from another, inventory management, invoice management, ordering, and analysis software from a myriad of other providers.

But is this really a sensible approach? If the whole idea is to centralize and simplify the telecom department, having all these new tools to manage may seem like more trouble than it's worth.

Some software firms have purchased smaller companies, and claim that the combined products will meet the complex needs of multinational corporations. The theory is that sewing up functionality from multiple applications would give greater overall benefits in a single application. However, as any good software architect will tell you, it is not that simple. Generally, it would actually be faster to rebuild applications from scratch than to try to cobble together pieces that were independently developed by completely different teams, built on different architectural foundations. The more complex the applications, the less likely the rate of success would be in trying to sew them up together.

There is no doubt that newer software will combine most or all of the major functionality that is currently available or promised today. This will be similar to Microsoft's Office suite, with applications that work closely together with each other. The best software will be built with a common architecture, and will have different modules available that provide customized functionality needed in complex installations. The only way to accomplish this is to carefully plan out the foundation before all the modules are developed, in anticipation of offering a suite of applications.

Within the next few years, dozens of these new firms that are not able to keep up with the market will either go out of business, or be acquired by larger companies. Eventually, there will only be a handful of major telecom management software providers left, and a number of smaller firms fighting for the leftovers.

The best way for a telecom manager to make the most use of technology today, without the fear of support coming to an end, or any other dead end that jeopardizes your investment, is to select your vendor or vendors very carefully. Evaluate the back end technology as much as possible, and ask probing questions about the architecture and future roadmap.

Is Telecom Expense Management Too Good To Be True?

The most common name for this category of software and outsourced services today is "telecom expense management." Search engines will point you towards dozens of companies selling services and software, claiming they will save you money and time. You may have spoken with vendors who sell "fully integrated best of breed, scalable, and robust software" or out-sourced services promising the complete management of your company's telecom assets and expenses. They promise a utopian environment where all your telecom troubles will vanish, saving you millions of dollars in the process.

Whether the offering is traditional software, web-based services, outsourcing, or anything in between, the rationale is that you can benefit from software tools to consolidate all your vendors' invoices into a single database, eliminate manual work, find billing errors, and easily analyze your costs.

It All Sounds Great...

... but here's the industry's dirty little secret. Like any other emerging technology, there are many bumps in the road. Most vendors try to gloss this over by promising every imaginable feature, in the hopes that you will sign a contract, wait for them to finish building their technology and services, and when they don't meet your expectations... as the saying goes, it's easier to ask for forgiveness later.

Like any other emerging technology that appears to have great value, and a solid market opportunity that will deliver untold riches, this niche attracts many pretenders. Some vendors have released press announcements and updated their websites with a full description of their telecom expense management software, but didn't mention that they hadn't even started building it!

If you've read any Dilbert comics, you'll get the impression that sales and marketing teams will do everything they can to sell a product, promising time machines and teleportation, and once they've found an unwitting customer, the pressure is on the engineering team to deliver. Some companies are more responsible than this, perhaps waiting until the product's concept is proven and development well underway, before rolling out the brochures.

Gartner Group's "Hype Cycle"

IT consulting firm Gartner (www.gartner.com) describes this phenomenon with their papers on what they call the "Hype Cycle." Briefly, Gartner's five stages are:
1. **Technology trigger – the breakthrough, product launch or other event that generates media interest,**
2. **Peak of inflated expectations – publicity generates over-enthusiasm and inflated expectations, with some successful technology applications but more failures,**
3. **Trough of disillusionment – failure to meet expectations makes the concept unfashionable, resulting in decreased media interest,**
4. **Slope of enlightenment – some businesses experiment with the technology to find practical applications,**
5. **Plateau of productivity – benefits become widely demonstrated and accepted; technology becomes more stable and evolves second and third generations.**

Regarding telecom management software, the signs are clear that we are currently in the second stage of Gartner's Hype Cycle – "Peak of inflated expectations." This is a signal for caution on the part of buyers. How do you ensure that your system becomes a "successful application," and not one of the more typical failures?

Top Promises of Telecom Cost Management Vendors

Complete automation of invoice information

Every carrier has different electronic bill formats, whether it's their proprietary invoice formats, PDFs, or even the "standard" Electronic Data Interchange (EDI) format. Although EDI is as close to a standard as it gets, only a certain number of vendors actually provide it, and when they do, it's always in a different variation. Any vendor that claims it's as simple as getting all your billing via EDI is either very optimistic, or downright deceitful. Realistically, you will be receiving electronic billing information in a multitude of formats, and your software provider will have to pick apart each one individually.

In many countries, electronic billing is available from all carriers, although still in different formats. In the USA, however, there are so many smaller carriers that there's a good chance your company cannot get electronic billing for all of your telecom services. Therefore, it's important to find out how much paper you will be stuck with, and how easy or difficult it will be to handle that load. There are some innovative ways to reduce data entry, but be prepared for some level of drudgery.

Whiz Bang Integration

An example of this hype is a vendor website with all the buzzwords that name every type of software that you could want to integrate with. The screenshot that goes along with this shows the Microsoft Internet Explorer window labeled "Save Web Page" that you get when you click on "File," then "Save As."

Some vendors offer a flat fee for each system to integrate with, regardless of the type of data or complexity of each system. A more realistic quote would be based on an analysis of the complexity and time involved. Some vendors may claim that their software will communicate with absolutely any other system, including your one-of-a-kind in-house application that they've never seen before.

The reality is that integration between disparate systems still, and probably always, will involve some programming work. There's no such thing as "plug and play" between systems that were not designed to the exact same standards.

Perfect Auditing

Some vendors promise that every phone bill will be cleansed, and any error will be automatically resolved. Right. It's true that most telecom carriers make numerous billing errors, and there is certainly a lot of money to be had in resolving them. However, out-of-the-box software cannot completely eliminate this problem for you.

Because telecom pricing contracts can be so complicated and different from one to the next, this functionality requires some customization. Even when you've got all the pricing parameters in the system, the invoice data will typically have some anomalies that are not necessarily errors, just odds and ends that were not accounted for or expected.

For instance, there may be special types of calls that are not part of any tariff or contract pricing, and after examining the details, you may discover that the pricing was actually correct. Software can point out anomalies, and if it's done well, this can help you find most (but not all) billing errors. People are still needed to review the anomalies and handle disputes with vendors.

Foolproof Inventory

On the subject of auditing for accurate pricing, one feature in most software is an inventory module that tracks what you know your company should be paying for, as opposed to what the vendors' invoices show you should be paying. Now, this doesn't mean that you will never be billed in error for a circuit that you've had disconnected, or that your inventory will stay up-to-date. If the system relies on human input, there will be human error. If somebody forgets to update the inventory now and then, it will quickly get out of date and inaccurate.

The only way to ensure that your inventory will remain accurate is to tie the ordering process and invoice verification to it. If you only place orders through the software, recording each cost item that's added or deleted, and your orders are tied directly to the inventory, you can keep your inventory up to date. If incoming invoices are automatically matched against the inventory on a line item level, you can catch the billing anomalies.

How Do You Avoid the Smoke and Mirrors?

Work with a vendor that will tell it like it is. If you're unsure, you can ask questions that are deliberately designed to gauge the honesty of the vendor that you're considering. If the salesperson that you're dealing with always says something to the effect of "Yes sir/ma'am, we can do that for you right away," then you have a pretty good indication of smoke.

A great example of a smoke-detection question, asked by a Fortune 500 technology company, "Do you support e-bonding for all carriers?" E-bonding is a term that describes having connections to carriers' back office software in order to streamline processes between carrier and client. Even more so than the mythical plug-and-play integration with desktop software, the promise sounds better than the reality. Each carrier has multiple versions of back office software for billing and orders, and most of it is prehistoric in computer terms. One news article reported that a large North American carrier has 360 billing systems, and that's just one vendor! Regardless of how well the software works, it would take decades and an army of programmers to try to sew up stable connections with every carrier.

Take Demos and Reports With a Grain of Salt

It's easy to demonstrate "vaporware" and make it seem like everything works. Be particularly aware of demos that only show screenshots, and not the actual application. Often, the screenshots for an application may look very appealing, however, sometimes this is done in order to make up for a mediocre technology engine. The truth is that the layout and graphics only represent a tiny portion of the overall application, but it's the only part that you can really see.

Many vendors focus on what report samples look like, which can be misleading. There are many more features to evaluate than just reports, but even within reporting there are more important factors to consider than how visually appealing they are. How difficult is it for a typical user to generate those reports? How long does it take for the software to create the report; a few seconds or a number of minutes? Can they be easily customized for your company?

Most importantly, what data lies beneath them? For instance, if you need reports on call details, make sure that the software actually stores that level of information. If you need powerful reports, you will need a powerful engine to create them.

Get Professional Help, and Look at the Blueprints

Whether you're buying an outsource service or software for your internal use, remember that you're buying technology tools, and you need to make sure that these tools are well built. If you're not a software architecture expert, be sure to get help from someone who is.

Most of the hardcore technology is under the hood, meaning that you would never see the engine, or the guts of the system, without opening it up. Ask to inspect the development documentation, or the "blueprints" of the software. Although quality documents are essential for planning and building complex software, you would be surprised how little of it most companies have.

Be sure to have the vendor's senior technology staff explain the documents and architecture. If you have chosen the right expert to assist you, he or she will be able to get an understanding of how sophisticated the application is, as well as the software development team behind it.

You Can't Always Rely on References

One large company purchased telecom cost management software after getting a positive reference from a fairly large bank. After signing the contract and having the implementation started, they realized that what was promised was far from what was being delivered. How is it possible that the reference was strong enough to make the purchase decision?

Perhaps the software was written specifically for the requirements of this bank, which turned out to be different than what the other company wanted. Maybe it's just that the people who made the decision at the bank to move forward couldn't admit to themselves that they made a bad choice. Or maybe it did work well at the bank, and for whatever reason, it just didn't work for the other company. It's also possible that there are close ties between people at the vendor and the client companies.

Ask for multiple references and ask some tough, in-depth questions, like:

- How many vendors is it handling?
- Is billing all electronic, and how detailed is it?
- How long was the implementation?
- Does the actual software match what you were promised? If not, what is different?
- Are there any "hiccups," and what are they?
- How many users do you have on it?
- Is it easy to use? Does it work quickly?

Be sure to ask detailed questions specific to the features that are most important to you.

Like any other kind of references, you are likely to be given contacts who are most favorable about the people involved, so take what they say with a grain of salt.

Since this is a new type of application, it's very likely that the vendors that you're evaluating may not yet have long-standing references. If you're considering being an early adopter, ask for a big discount and favorable terms in the agreement.

Perform a Thorough Test Drive

Whether you have references or not, you need to know that what you're buying will perform to your expectations. You wouldn't buy a professional race car by letting the salesperson talk you into it, or watching somebody else drive it, would you? Similarly, a drive around the block wouldn't be enough to convince you.

If you have dozens of vendors, thousands of invoices and inventory items, and hundreds of millions of call details, make sure that the pilot project you do contains enough data so that it's comparable. Poorly designed software will slow down dramatically, or even crash, with larger volumes.

Get a Solid Guarantee

If there are certain criteria or functionality that you absolutely must have, ask for something in writing. Make sure that your contract allows you to terminate it (and get your money back) if your specified deliverables are not met.

Who Should You Consider Buying From?

When the value and opportunity of an emerging market become clear, new product and service offerings will proliferate through new companies entering the market, and participation by other companies branching into the market. This is not always a bad thing, as every product or service was new once, and this can lend vitality to the market.

However, it is important to be cautious in any purchasing decision, particularly with regards to:

New Companies

It's amazing just how many new companies did appear once analysts and media started talking about the size of the market. Some of these companies disappeared within a couple of years.

Why the eagerness to enter the market? To some business people outside of the industry, the problem of telecom invoices and costs may not seem all that difficult to solve by hiring programmers. That is, until they get knee-deep into the intricacies, exceptions, ambiguities, and odds and ends that are discovered along the way.

If a company is well-funded with millions of dollars in venture capital, it's easier to spend money on marketing and sales, instead of developing a top quality product. After all, the business theory is that a company with the best marketing, even if it has only a mediocre product, will beat out the company with the best product if it is not promoted effectively.

Software that's not from a software company

Software development is a rigorous discipline, and even most dedicated software development companies aren't all that good at it. If you work in a large company, you've probably witnessed attempts to build various applications in-house because it seemed cheaper than paying license fees to specialized software vendors. Back in the days when corporate finance software was an early industry, many companies tried to build their own ERP and finance software, yet today, almost every major corporation uses software from one of a handful of proven vendors.

Several companies that provide other types of services such as invoice processing, Move/Add/Change orders, telecom audits, and various flavors of outsourcing decided to build software in-house. Unlike a dedicated software development company, they either hired a bunch of programmers or outsourced the development and hoped for the best.

Software Companies from a Different Business

Even though they may have a decent development team

and history, if the company has traditionally focused on something other than telecom expense management, chances are that the product will not be world class. Many call accounting companies also jumped on the bandwagon, figuring that invoices, inventory, and orders aren't that much different than their traditional business. The smarter companies that want to enter the market partner with the best company they can find, instead of trying to re-invent themselves.

OK, So What Kind of Company Is Left?

The best software providers will have both a full in-house software development team, and the telecom cost management industry expertise developed over numerous years. Like any other industry, the best products and services tend to come from companies that have a singular focus on a particular goal.

As an example, Avema Corporation has been providing telecom audit and cost reduction services since 1995, and although we recognized the software opportunity in 1999, it wasn't until 2001 that we were able to change the focus of the entire company to development of world-class telecom expense management software. Before then, we tried hiring contractors, outsourcing software development, and other mistakes that resulted in a mediocre product before realizing that the only way to develop a top notch application is to invest all of our collective efforts and resources into it. Our early prototype never developed into what we expected, and if we had led clients to believe that it was already built, it would have been very difficult to recover.

Since then, every employee's focus is now on software, and our elite development team has built highly sophisticated technology that we are proud to supply to our clients.

Chapter 11

Buyer's Guide

What are the most important specific criteria
for your organization when buying software tools
for telecom management?

What Are Your Requirements?

It's difficult to figure out everything up front about what you
will need, but it is certainly worth spending the time to plan as
much as you can. Start by asking questions like:

- How much does my organization spend in total on
  telecom?
- How many vendors? How many invoices?
- What currently already works well, and is there room
  for improvement?
- What does not yet work as well as I would like it to?
- What are my biggest challenges or headaches?
- Is it just for the telecom department, or can other
  business units make use of these applications?
- What is the long-term scope being managed by the
  telecom department? Are the responsibilities expand-
  ing to include other geographic locations, or is the

company likely to merge with another?
- What are the most important functionalities that will save the most time in my department?

Who Are You Buying From?

It is important to understand the core competencies and focus of the companies that you consider buying from. Many vendors with backgrounds in non-telecom areas look at what seems to be a burgeoning market, and decide that they want some of that business. However, it is important to be sure that the people involved in the work have a background in telecom management. It is easier to take that knowledge and apply it to software development, than it is to take a software designer and teach her or him about telecom issues.

In evaluating software, consider the amount of experience that the company has in this field, and the breadth of its offerings. This gives you an idea of its commitment to the telecom sector, and indicates whether you can count on its being there for you when problems develop, viruses strike and upgrades are needed.

Usability

One of the key factors to consider is how easy the application is to use. The ideal software is powerful, yet easy for anybody to understand and use. Of course, you must be able to perform all the functions that you need to, but you also have to be able to quickly find the buttons or icons to run the program. More importantly, your end users who may use the software for retrieving reports or drilling down into invoice details, must be able to do this without significant training, if any. From the point of view of professional software and user interface design, this is actually very difficult to accomplish. For this reason, you will find that many applications have far more buttons and things to click on than necessary. You may also notice that different styles of navigation, buttons, links, etc. are used inconsistently on different pages.

Consider an analogy of driving your car. If cars had all the dials and buttons in a 747 jet cockpit, there would be many more accidents. Fortunately, there are only a few controls that you need to use and be aware of in order to be a successful and safe driver.

In North America, all car rentals from major companies are automatic, even though the cost to purchase automatic cars is higher than manual ones. Since some people who rent cars are not regular drivers, having fewer controls to maneuver makes it less likely that they will experience any problems. On the other hand, most car rentals in the UK are manual, and it costs more to rent an automatic. Between driving on the opposite side of the street, sitting on the opposite side of the car, and shifting manually with the opposite hand (although the pedals are the same), it's surprising that anybody from other countries rents cars there at all!

Some of the most common criticisms from would-be software buyers that I've heard about various applications, are that they're either too complicated to use, with too many buttons, or they simply don't do what they want it to do.

Think of the Lowest Common Denominator

Just because you or someone on your team may be a computer whiz who can run obscure software and find all the hidden buttons and menus, it doesn't mean that you can skip an evaluation of usability.

Who will be using these software tools? Obviously, the telecom department will make extensive use of it. In a smaller company, if you're the only person who will ever use it, then the decision becomes much simpler. However, in most companies there will be multiple users, including many employees outside of the telecom department. Even if you're the only one who will initially be using the applications, you may still want to consider having someone else be able to fill in for you when you're away, or when you move on to another position (after you implement everything in this book and get a promotion!).

Staff in the finance department are the next most likely to be using at least some of the software functionality. They will require access to reporting, and the ability to create ad hoc queries to find specific information. Again, the software must have the power and flexibility to allow you to accomplish this, but easy enough to use that a novice computer user can get the hang of it.

Need for Speed

In the same category as usability, speed is of the utmost importance. Can you imagine waiting for several seconds in between every single page view, or waiting several minutes for a simple report to be created? Especially in today's computer culture, we've become accustomed to quick responses. Have you ever waited for a computer application to start, and by the time it's open, you've forgotten why you had started it in the first place? Imagine this happening every day for every user with your telecom expense management software.

Speed is particularly important for larger installations. If you have hundreds of millions of call details, or tens of thousands of charge items, a poorly architected software solution may crawl agonizingly slow, or even crash, even though it may work fine with a smaller amount of data.

Other Issues to Consider

How Much Data Can It Handle?

Typically, there is a direct correlation between speed and the amount of data in the application. Also important to consider is how much detail, and at what level, is it available? For example, if you find out that the recurring monthly charges have increased, it doesn't help much if you can't see what caused that change. Similarly, if a branch office's long distance costs suddenly spiked upwards, the manager may need to see the individual call details to determine exactly what caused the spike, if it was a legitimate reason, and if it is expected to continue.

If call details from the invoices are important to you, whether to have the rates verified, or to be able to view them, consider how these details are handled by the software. Compared to the number of monthly recurring cost items, there could be hundreds of times more call details to store in the application database. For some of the larger organizations, this could be hundreds of millions of calls per month. Since speed and performance are directly related to the amount of data and how well the system is architected, only the most high end software will be suitable for larger organizations.

How Flexible Is the Software?

Many commercial applications, especially complex ones, are initially designed for one client. This could mean that many functions and setup procedures are hard-coded to accommodate that one client's specific requirements. Customizing the same software to your environment could prove to be difficult, or even impossible.

A simple example would be company hierarchy. The company that the software was first built for may have been structured by site and location. If the software was designed only to accommodate these two levels of hierarchy, significant workarounds would have to be done in order to customize it for your company's use.

A more complex example could be in synchronizing your list of employees from your Human Resources department to your telecom expense management's directory module. In order to do this in an efficient way, each employee must be mapped from your HR department's feed to the directory. If the directory's structure isn't designed in a way that enables you to match the exact structure of your HR feed, then you may not be able to use this useful feature.

Make sure to consider your long-term requirements, and further customization that you may want. Can reports specific to your company be easily created if necessary? What other needs will you have in the future?

In order to accommodate the unique needs of different customers, the software tools must be designed with this flexibility from the ground up.

Permissions for Functionality and Data

In order to make full use of telecom expense management software in a large organization, it must be designed for access by multiple users with different job functions, and different needs within the software. Otherwise, the flow of information would be stifled, since only a few telecom managers could access the data.

If the CFO requires access to view high-level reports, but

doesn't want to be bothered with the nitty-gritty of your data network inventory, you must be able to set his or her permissions within the software to only allow access to the relevant data. Similarly, an accounts payable clerk may only be authorized to view invoices, and nothing else.

Still more important is being able to limit access to specific sets of data. For example, a business unit manager may only be required to view costs that are allocated to that business unit, but should be forbidden access to any other data. This means that not only should his access be limited to reporting, but it should only show him the subsection of that data that relates to his unit.

Web-Based vs. Client/Server

Before 2001, most enterprise software was written for a central server, and designed to be accessed on the company network. To extend accessibility beyond the local network, it had to be reached through expensive wide area network circuits, which was also complicated to set up.

Today, most applications are web-based and are accessible from any internet-connected computer through a standard web browser. If you need to have people access the application from multiple locations, this is very important. If you never plan to have anyone outside your department or site use the software, then either web-based or client-server would work.

Multiple Currencies

For example, an application designed for an American company to handle only their American telecom bills will not be able to accommodate multiple languages or currencies for a global company.

Some software companies may try to add multi-currency functionality as a "feature" to an existing system, however, this would be a painful error. What is involved in consolidating billing in multiple currencies into one? Although it could seem fairly simple on the surface, consider the complexity of what's involved. Throughout the entire application, every place where there is a currency amount, the type of currency must then be included as well. If you want to be able to view what the currency exchange was in an earlier period, the system will also have to store the historical exchange rates for every currency amount in your entire set of data.

Trying to patch this type of complex functionality on as an afterthought in the software design can lead to errors in your currency figures. It would be too easy to miss detailed lines of code that need to be updated. What's worse, you may not always know about these errors, since you wouldn't necessarily get an error message, just an incorrect currency amount.

Multiple Languages

For international companies, this functionality would allow your staff who are more comfortable with languages other than English to use the software just as easily. Of course, if your company's operations are all in English-speaking countries, then it isn't necessary. Similar to having multiple currencies, this functionality must be embedded in the foundation of the software.

Chapter 12

Functions of Telecom Management Software

Call Accounting

This technology has been around for decades, and many companies have made good use of it. However, not every company has been able to extract the full business value from the reams of data that this can produce.

Invoice Processing

It would be hard to imagine telecom expense management software that does not handle invoices. If the software is designed well, this may be the area where the most automation can occur.

Loading

Ideally, all invoices would be available in electronic formats, however, in the United States, there are many smaller carriers that can not easily provide this. Even when bills are electronic, they are in many different formats. Sometimes, there may be several bill formats from the same carrier, and these formats may change on a whim.

This makes it very important that the software vendor continuously updates the data translators that import the invoices. Ideally, the vendor should import them for you, and verify the data for accuracy.

For the paper bills that you still have to deal with, the software should make it easier for you by automatically filling in fields from the previous month's invoice when you select the account. In this way, you can minimize the time spent on inputting paper bills. It would be unproductive to type in call details, however, you could still benefit from having individual cost items entered, depending on the amount of data involved.

Presentation

The major purpose of importing the invoices is to have a central repository of data. Although you should be able to perform analytics on the data through reports and alerts, you will also sometimes want to view the invoice data itself. Having all the invoice information available and easily accessible becomes important for this.

Allocations

Since this is traditionally one of the most time-consuming tasks to perform manually, here is an area where software can make a big difference. By mapping the invoice costs to your General Ledger codes once, the software should be able to copy the same mapping from month to month, eliminating any work on your part other than handling exceptions. Sophisticated software should allow you to allocate the entire account, subaccounts, individual phone numbers, or monthly cost items. Since some of these will need to be allocated across several GL codes, there should also be functionality to split costs and assign the proportions by percentage.

Auditing

There are several ways to verify costs on your telecom invoices. Individual call details can be verified if the software stores the contract and/or tariff rates and is programmed to compare these rates to the actual pricing on the invoice.

In addition to the rate verification, some software will notify you of any sudden increase in costs or usage. This can be used as another layer of verification to make sure that billing errors do not slip through.

For monthly cost items, invoices must be reconciled with the inventory.

Inventory

Your inventory is a list of every monthly recurring charge that you are expecting to pay for. This should include the invoice account and subaccount information, phone or circuit numbers, and individual charge items.

Reconciling invoices with your inventory can eliminate billing errors that occur when the vendor "forgets" to remove cancelled circuits, or when new charges appear on your bills mysteriously. The only way to streamline this reconciliation, and avoid doing it manually, is to mirror the invoices with the inventory, so that each item from the inventory corresponds with an item on the invoice. This means that the inventory should initially be created from the actual invoices, and be continuously updated based on new information from invoices.

You may already have some sort of inventory on spreadsheets, however, even if you were able to import it to the software, the only way to perform automatic reconciliations would be to match up every single inventory item to the corresponding invoice item.

If the inventory must be updated manually, chances are that in a large organization, people will sometimes miss this step. Eventually, this could mean that your inventory becomes useless the further it slips from being 100% accurate. There is no guarantee that the process will be foolproof, unless ordering is also tied in.

Ordering

By placing orders through the same application, and tying orders directly with the inventory, you can eliminate any manual updates to your inventory. For example, if you cancel a circuit and its associated cost components, you should be able to mark them as cancelled in your inventory. When the invoice for that account is imported and the reconciliation takes place, the software can then notify you if the vendor has not yet removed the charges from your invoice.

When new items appear on the invoice, the reconciliation process can identify them, and let you decide whether or not they are legitimate charges that should be added to your inventory.

Reports

All telecom expense management software should include some kind of reporting. Basic reports will help you analyze your costs and usage. Capabilities like being able to perform an ad hoc query at any time can give you a lot more power. Well-designed software, including reports, will let you select the specific information that you want to see, and make it easy for you to access and understand it.

Some advanced functionality may include allowing you to set different parameters on a given report, and the ability to save that exact detailed view. The ability to schedule and automatically e-mail specific reports to individuals can save you time. For example, once a month you could send to each manager a report showing his or her cost trends for the most recent three months.

Alerts

Still more powerful than reporting are alerts that let you proactively manage exceptions. Notices can be e-mailed whenever criteria that you've predefined occur.

Instead of sending out a report every month to a business unit manager, you could send an e-mail only when the costs increase past a certain threshold. This way, the manager would only need to react when there is an exception, instead of just scanning a monthly report.

Some other uses for alerts include:

- Notifying the accounts payable department when new invoices arrive and are ready to be paid.
- Monitoring usage policies, such as wireless phone use.
- Warning of an approaching due date, in order to avoid late payment fees.
- Proactive notification of errors on the invoices.
- Notification of upcoming contract due dates.

- Calls being made to certain countries or phone numbers.

Alerts can really help you avoid routine administrative tasks, only requiring you to respond when an exception occurs. Working this way can free your time up more than anything else.

Directory

A directory is simply a list of employees in the company, associated with the various services and devices that are assigned to each person. This is useful to track who has what at any given time. It is particularly useful when people leave the company, to make sure that all equipment is returned, and all services are either cancelled or assigned to someone else. Similarly, this makes it easier to assign services and equipment to new employees. If you want or need to report costs by employee, some type of directory will be essential.

In larger companies, it may be helpful to synchronize your directory with your Human Resources list of employees. Instead of having two separate lists of all employees, this can reduce potential for errors.

Disputes

Since telco billing errors are inevitable, you will be disputing them on a regular basis. If you find that you have many disputes at any given time, it can be helpful to have a way to track them. This software functionality can help you organize what items are being disputed, the amounts, and any comments.

Wireless

Because wireless devices are scattered throughout the organization, and each one has different usage, features, and even a different pricing plan, there are some aspects of telemanagement software that are specific to managing wireless costs.

If your company decides to manage price plans, and continuously optimize pricing by choosing the best plan for each user, you may need software for this.

Section IV

Trends and Conclusions

The only consistent thing about telecommunications is that it is constantly changing and evolving. There is no way to be prepared for all of the possible changes to the environment over the next five years, let alone ten. Ten years ago, would you have imagined thumb-operated converged wireless devices like a Blackberry or Treo, or that IP telephone systems would become a standard technology?

Let's consider current trends and take a look at where we're going in the future.

Chapter 13

VoIP Will Take Over the World

VoIP is starting to create a wave of change in the telecom world. How can you take this opportunity to improve your career?

Attitudes About VoIP

What's the current prognosis regarding VoIP, the technology that has telecom managers apprehensive, excited, and mystified all at once?

In the Avema survey, most acknowledged that the evolution to VoIP is inevitable, but were not enthusiastic about this change. Most telecom managers who handle voice, but not data communications, are extremely cautious about VoIP and IP telephony. Voice managers are justifiably proud of the almost-always-on capability of their systems, and almost bombproof reliability even in the face of power outages. They also like the relative invulnerability of the traditional telephony system to worms, viruses, hackers, deniers-of-service and other nightmares facing computer systems. One telecom manager, Lana Norkoski, says "As a hospital, the idea of using VoIP scares me." This is very understandable, given that lives may be at stake if there is any downtime. For a hospital, the difference between 99.99% and 99.999% uptime can be a lot.

Many voice professionals don't like the idea of their organization's phone systems being entrusted to what they see as the lesser reliability they have seen in the computer networks. Also, for many, stepping outside of a known and comfortable operating environment can be intimidating.

As you can imagine, managers who handle data communications tend to be more optimistic about VoIP.

Converging the telecom department with the IT department is yet another challenge, since the two have traditionally operated in very different worlds. Both sides need to have a much more thorough understanding of the other's environment and culture before even considering a major change to VoIP.

What's the Verdict?

To date, it is difficult to get a comprehensive overall view of the benefits and disadvantages of VoIP. There are so many articles and studies published that give different, often conflicting, statistics about the rate of adoption and how VoIP is being used in many companies. The one thing that seems clear to just about everybody, is that VoIP is here to stay, and is without a doubt the technology that we'll all be moving towards. Of course, everyone has a different opinion and sometimes a bias, but here is what the consensus amongst many managers, vendors, and analysts indicates:

The main reason for migrating to VoIP used to be cost savings on long distance, when rates were higher, and dramatic reductions could be seen by transporting intra-company calls over the existing data network, instead of paying carriers by the minute. Considering the fact that carrier long distance pricing has dropped so dramatically, this area for cost savings becomes more and more difficult to justify. Many carriers are already using VoIP on their own dedicated networks to carry long distance traffic, and are basically performing the same functions as a large company voice network, but with greater cost efficiency because of their scale and volume. To get the optimal voice quality, either network would require the same amount of bandwidth for each call. However, since the carriers are doing this for millions of customers, their cost to run the network will be significantly less than for a single company, even a very large one. The exception to this may be international calling, if the per minute rates are still relatively high to other countries where you have offices.

If your organization has dedicated circuits for both voice and data between many sites, it is more likely that cost savings will be justifiable. Of course, you may also save money by eliminating the dedicated voice circuits in favor of per minute long distance with your carrier. On the other hand, if most of your voice communications are with people outside your organization, then there may not be any cost savings at all in this area. If your company is upgrading equipment specifically for this purpose, it may actually cost you more than you'll save.

A more tangible cost savings comes from quicker MACs. Nemertes Research calculates that a typical MAC on a TDM system costs anywhere from $65 to $350 when outsourced, with an average of $105. MACs performed by internal staff on a TDM PBX cost an average of $33. With an IP PBX, however, the average cost is only $5 per MAC. End user employees could even perform their own MACs in routine cases.

Aside from cost savings, other benefits include added features and functionality that are only possible, or that are more readily available, with VoIP:

- Unified Messaging – Having voicemail, e-mail, and faxes all stored in one place can increase productivity by making all this information more easily available, and in different formats. For example, some people like the idea of being able to "listen" to their e-mails by having it read to them over the telephone. More common would be the ability to access voice mail through any computer with an Internet browser.

- Greater support for remote users and branch offices—The IP PBX can be extended to home offices and other branches, meaning that these workers outside of the main building can use the same features, without requiring another telephone system.

- Communications collaboration—Applications like sharing documents on a PC and desktop videoconferencing over IP open up whole new possibilities for more efficient communications.

- Follow me, find me—When calls come in, the IP system can be programmed to ring landline and mobile phones in different locations, either simultaneously or in sequence, increasing the accessibility of your end users, and hopefully eliminating a lot of voice mail tag.

- Other new features and applications that are only imagined because of IP, such as how the town of Herndon, VA is broadcasting Amber Alerts to the municipality's entire workforce—meaning hundreds more pairs of eyes can be on the lookout if a child is abducted. There may be a whole slew of new applications developed over the years that are only emerging or haven't even been thought of yet.

Although there are numerous benefits, there are still reasons for some concern, including:

- Overall reliability—It can take longer to troubleshoot because the IP system relies on the company's data network and all the components that it is made up of. 99.999% uptime is more difficult to achieve and requires more redundant equipment, which affects the value proposition. What happens when a softphone is unavailable because the computer is in the middle of a reboot or crash? Some VoIP installations may mean more down time for phone users than they are used to.

- E-911—With PBX-based systems in buildings such as a university dormitory, emergency workers will know from which location an emergency call originated, allowing response times that can be life-saving. Since IP phones can be moved to different rooms, and even different buildings or cities, these issues need to be addressed. There are numerous solutions being worked on, but so far the consensus is that it's not quite ready yet.

- Security and spam—In the 1970s, some telephony enthusiasts discovered the joys of "free" long distance. Called "Phone Phreaks," hackers famously learned to duplicate DTMF tones with low technology – a toy whistle available as a premium in boxes of Cap'n Crunch cereal that reproduced the 2600 hertz tone needed to authorize a call on AT&T's network. What will today's hackers, with all the free-on-the-Web technology available to them, be able to do if they are determined to hack your organization's VoIP network? If you thought that unwanted telemarketing calls are a nuisance, what will happen when spammers are able to access your system and send cut-rate pharma offers to all employees, several times a day? Like any other network, there are precautions that need to be taken to maximize security.

- Limited amount of expertise available—At present, there doesn't seem to be enough installers and staffers that are familiar enough with both the voice and data worlds, enough that you can be assured of a smooth transition to VoIP. Expect a learning curve, and do what you can to minimize it through training and being selective in whom you work with.

- Compatibility with still evolving standards— Although most of the technical standards for operating VoIP are reasonably established, there are still some that are in flux. This probably wouldn't mean that you would have to rip out your shiny new IP PBX in the near future, but it may mean that you will have to keep up-to-date on new standards as they emerge, and spend time and energy migrating towards them.

The issues described are very real, but they're temporary. They're all being addressed, and at some point VoIP will become the new standard.

In 2005, we are still witnessing the adolescence of VoIP in the marketplace, although it seems to be growing up very quickly. Few organizations have installed a significant VoIP capability, and fewer yet have converted their entire operations over. There have been some high-profile projects over the past few years, and we are only recently at the point where the majority are planning to make the shift in technology for their future equipment needs. A 2004 Yankee Group survey found that 39% were either testing IP telephony or had partial installations, whereas only 5% were fully deployed. Nemertes Research conducted a study of IT executives, and found that 71% were either piloting or had deployed VoIP, 25% were planning to, and only 4% did not intend to use it at all.

How Does VoIP Affect the Value And Career of a Voice Professional?

A *Network World* article from November, 2003 states:

> Once IP telephony is implemented, many companies hand off MACs to help-desk support or other IT staff, which reduces the total time needed to about 15 minutes. In that case, per-MAC costs drop to between $9.25 and $11.25 per hour. Because those highly trained voice technicians are no longer needed for MACs, companies can opt to reduce headcount or reassign those staff members to other areas.

If you manage voice communications, but not data, this is particularly important. With thinking like this, if the executives are not aware of what the voice department does and the value that you bring, they're only looking at how much your salary costs the company. A major motivation for VoIP is "reduced operational costs," which can almost be translated to "Hey, you can get rid of your voice communications managers" because your IT staff can now handle voice just as easily. They've been managing the computers all this time, how hard can it be to have them do the phones?

From a shortsighted CIO and CFO point of view, this sounds great. They figure that they can get kudos for getting rid of unnecessary expenses, i.e. your salary, and maybe even get themselves a bonus.

Of course, as voice professionals can predict, the reality will be quite different. Traditionally, voice and data departments have been separate in most companies, and with all the talk about convergence, it's been challenging to figure out who should do what.

Robin Gareiss, Executive Vice President at technology analyst firm Nemertes Research, says that attempts at having both voice and data staff share leadership in VoIP implementations tend to "end in disaster."

She says, "I've seen the most successful projects where there isn't this infighting between the voice and data people going on. You've got to make a decision, as a CIO, to put one or the other in charge, but not both. If you've got a person in the organization who has an understanding of both voice and data, that's the person you want in charge of the project, but most companies don't have that."

Gareiss declares that there is an urgent need for a new type of specialist who has a thorough understanding of both the voice environment as well as the data network. She says, "These experts need to understand how to manage a network that contains both real-time traffic, traffic where it's acceptable to have delays; all the different traffic types. How do you effectively manage all that in one network infrastructure?" Gareiss also suggests that these specialists will command a premium over someone who is only familiar with either voice or data.

The good news for voice professionals, is that executives are more likely to appoint employees with a voice background to lead VoIP initiatives. In a 2004 Nemertes survey, 38% of companies appointed staff with a voice background to lead the VoIP implementation, 21% appointed the data staff, and 41% both. In 2005, however, 43% of projects were led by voice staff, 31% by data, and 26% both.

Nemertes recommends that someone from the voice side is put in charge of VoIP projects, and ensure co-operation from the data staff. "If your CIO empowers you as a telecom person to run this project, take that empowerment and run with it. Find the data people in the organization who are going to work with you, and not against you."

Gareiss provides some reasons why organizations should appoint voice staff to lead VoIP initiatives, Voice staff have:

- a better understanding of voice as an application and the network performance requirements,
- a better ability to help the organization through the transition, particularly in a hybrid environment with both TDM and IP PBXs, and
- more likely to focus on maintaining the TDM PBXs' uptime as a mission critical application during the transition.

However, she cautions that voice professionals are in a "do or die" situation. "If you just have a telecom background, you'll be valuable for a few more years, but as everything does merge onto the network, and the data people do start learning a little bit about voice traffic, your value becomes less and less." Data networking specialists do not have this same level of motivation to learn voice technologies.

The more proactive voice professionals have significantly more opportunity ahead. As Gareiss comments, "Telecom people are in a great spot! If you're a data person, where do you go to learn telecom these days? If you're someone with a telecom background, you have knowledge that is getting more and more rare. It's much easier, much more pervasive to get training on data networking or network management, than it is to learn about voice."

Chapter 14

Outsourcing

What would happen if your department got outsourced? How will you ensure that you will always be of value to your organization?

Telecom Outsourcing

More and more vendor companies have been calling us at Avema, talking about providing telecom outsourcing, and not just bill auditing or invoice processing, but their goals are to eventually provide full telecom department functionality. To make their operations as lean as possible, they are asking us about how to automate their internal processes. These aren't new companies springing up either, they are established equipment distributors, service resellers, and service firms that see the opportunity for a bigger role in the market.

Now this makes tremendous sense for clients who are small enough that they don't have a full-time telecom manager or department, and that's probably where this trend will start. Companies can pay for the equivalent of a quarter or half of a telecom manager, or supplement their internal staff. But there are larger outsourcing deals happening as well. Bell Canada has bought several outsourcing/consulting firms in the last year, and they are targeting some of the largest companies in the country as customers. Telcos in the US are looking at telecom expense management software as something that they might provide to their enterprise clients, and some US telcos are already actively involved in outsourcing.

However, in a large company with a dedicated telecom department, if it's run as efficiently as can be, and their results are being reported and showing solid ROI, there is no reason to consider outsourcing the telecom department by itself. Even if the corporate decision is to outsource as much as possible, at least the telecom personnel will likely continue to be employed by the outsourcer.

IT and Back Office Outsourcing

In 2004, Procter and Gamble was taking bids to outsource every part of their back office functions, basically everything that was not considered a core competency in the company. This may become increasingly more common.

Outsourcing of IT functions has been accelerating dramatically for the past dozen or so years, although there is currently a lull in the market. As more experience has been built up with out-sourcing, companies are learning more about best practices, how best to manage outsourcing contracts, and generally learning from outsourcing deals that went badly in the past. Historically, companies jumped towards the opportunity to outsource largely for cost cutting reasons, or to "make this headache go away," as Robert Eveleigh, AVP Telecommunications for a major insurance company, says. Eveleigh describes how most outsourcing in the past was an attempt to fix a broken problem by getting rid of it, rather than fixing the problem internally first, and then using outsourcing to improve upon it. Otherwise, there is no way to measure any benefits of outsourcing, or to control it.

With VoIP, voice communications will eventually become just another application on the data network, which means that it will fully be a part of the IT department. If the trend towards outsourcing IT continues, it will become more likely that the converged telecom department will be outsourced as well.

In every outsourcing deal, some staff need to be kept as internal employees in order to manage the outsourcer. More senior positions in the company are much less likely to be outsourced.

Offshoring to Lower-Wage Countries

Who would have thought that security cameras located in a European shopping center could be monitored, remotely, by personnel in Africa? Or the phenomenon of "medical tourism," in which an American heart by-pass patient can fly to India for an operation that is done to First World standards, yet even with the travel expenses costs less than it would be in the US, with a visit to the Taj Mahal included?

Anticipate that if something can be done effectively in a low-wage country, it will. But also assume that the technology will allow more work to be sent offshore, and rising education and skills in low-wage countries will mean that more and more work can be done overseas. Even now, an army of fully-qualified CPAs in India is able to prepare American citizens' income tax returns for a Big Four accounting firm.

So, don't try to hold back the tide. Be prepared to part with any functions that can be offshored. Better yet, analyze the factors and determine which functions should be offshored now, which are candidates for offshoring in the next few years, and which should stay close to home for the foreseeable future.

For example, straightforward help desk support can often be effectively done by trained staff in countries such as India. However, more complex tasks such as trouble-shooting complex programs may be best retained in the home country. Analyze whether the cost savings of a move offshore will be worth the risks – including sometimes-dodgy systems reliability in developing nations.

Seize the future by analyzing which functions are likely to be offshored, and make sure your job description doesn't include them. Look for ways that you are in a strong position to add value, for reasons that can't be duplicated by someone in a low-wage country. Don't rely on paper qualifications – universities in places like India and China are graduating large numbers of well-trained, skilled individuals every year.

One of the most sustainable advantages you can have is the intimate knowledge you have of the organization, its customers, competitors, and the culture in which it works.

Chapter 15

More Long-Term Trends

*What changes will happen in the telecom market
within the next ten years? How will this affect you?*

Centralization and Consolidation

Is Your Telecom Department Going Global?

Many multinational corporations have spent the last few
years wringing out savings from their US telecom invoices.
Since the results have shown such a dramatic amount of
refunds and cost reductions, the natural next step is to extend
this cost reduction work outside of American borders into
other countries where these companies operate.

Going beyond bill audits and contract negotiations, many
companies have also decided that once they have cleaned up
their bills, they want to continue to keep them clean. Otherwise,
it would be another massive undertaking a couple of years
later to do an international telecom audit.

With the newer web-based management tools available,
they are planning for the future, and keeping a central record
of all their circuit and charges inventory globally. With the
right processes and software, each region can still pay bills
locally, and have access to all the cost and usage information

within their regions, in their native language. The central telecom team can access all the information from invoices and inventory around the world. Reporting can also show consolidated billing, converted into the company's main currency. Having all this data stored in one place, accessible through any Internet browser, makes this possible.

In other large companies that are not necessarily spread across multiple countries, this same trend towards centralization is occurring. Where there are multiple independent divisions of a corporation, the telecom contracts are more likely to be negotiated centrally today. Consolidating billing and centralizing bill processing is also more common than before.

Bandwidth Gets Faster and Cheaper Every Day

If you're like me, you may remember experimenting with dial-up modems long before the Internet was the next big thing. In the early 90s, 2400 bits per second was considered fast. Compared to that, dial-up speed of 56,000 bits per second was considered the "on ramp to the information superhighway" towards the mid-90s. Multiply that by a further 20 times, and we have today's standard Internet access available in just about any neighborhood in most of the developed world. This all happened in a span of about ten years.

A non-profit organization, Internet2, comprised of over 200 universities, some private companies, and government, has already been working on a much higher speed private network to simulate future Internet applications. They call the network Abilene, and it currently has bandwidth of 10 Gigabits per second. For the mathematically challenged, that is *ten thousand times faster* than today's residential broadband access. That's over *four million times faster* than my clodgy old modem from the early 90s.

Is it only a matter of time before this type of bandwidth is available to every organization, or even every household?

Why pay millions per year for a private WAN when this kind of Internet access is readily available to anyone? There is already software available for networking sites over the public Internet, and this trend can only accelerate as more proven solutions concerning security are rolled out, and organizations become more comfortable with them.

All that will be needed is one Internet connection at each site, which may only cost $50, or a few hundred dollars for a larger site. Even with two separate providers to provide redundancy, it will be much less expensive than what we pay today.

Wireless Proliferation

A recent international study by Sadie Plant of the Cybernetic Culture Research Unit at Warwick University in the U.K. tested children to see how they would ring a doorbell. While earlier generations would use an index finger, many young children preferred to use a thumb. Why? Growing up on thumb-operated gaming devices, it has become the digit of choice for many tasks. The rise of thumb-operated communications devices can only accelerate this trend.

Many young people are highly adept at thumbing their way through a message on mobile devices such as a Blackberry. As younger people continue to enter the workforce, you will need to deal with a world where more and more people take wireless connectivity for granted.

Several hardware vendors have announced development of wireless phones that combine both Wi-Fi and GSM or TDMA "cellular" technology. These will be able to roam between networks, so that when you have Internet access, you will not need to pay wireless carrier fees. This will reduce wireless carrier bills, since today many cell phone calls are actually placed near employees' offices. Roaming and long distance usage will also diminish. We will depend less and less on the traditional wireless carriers, and eventually they may become largely irrelevant.

Losing It

Have you ever forgotten your mobile phone or PDA somewhere? Hopefully, you were able to quickly retrieve it, but it's not always possible.

A study from January 2005, sponsored by Pointsec Mobile Technologies, showed that mobile device users worldwide tend to leave their mobile phones, PDAs, and even laptops in taxi cabs. The highest rate of lost devices per taxi was in the USA, where a Chicago taxi company was surveyed. In a period of only six months, over 85,000 mobile phones were left behind in Chicago taxis. Additionally, more than 21,000 PDAs were left, and over 4,000 laptops.

This study also uncovered other interesting items left behind, including a harp, a throne, a cat, a treasure bond worth $2.5 million, a prosthetic leg, and a baby.

What is also interesting is that the number of devices lost is increasing dramatically every year. Considering the overall growth of the market, and the shrinking of the actual devices, this is understandable.

As we become more and more accustomed to our mobile devices, storing ever larger amounts of data on them, it is becoming increasingly important to protect this data from random strangers that may find them. Password protection or even biometric recognition, as well as encryption of data, will play an ever more important role.

Wi-Fi, Wi-Fi, Everywhere

Several cities are proclaiming themselves to be a "Wi-Fi city," including Grand Haven, MI, Half Moon Bay, CA, Pune, Jerusalem, Zamora, and Amsterdam. This means that wireless Internet access is available everywhere in the core of the city, indoors and out. Many other cities are preparing to launch similar initiatives.

Spokane, WA is offering this Wi-Fi access free to everybody for up to two hours per day. The city, partnering with private companies, helped to build the network for its own use. Mobile city employees are using the wi-fi access in their daily routines, such as police officers running license plate checks, issuing parking tickets, and more.

According to Spokane city officials, the technology used only cost between $50,000 to $75,000. With this kind of ubiquitous and inexpensive Internet access beginning to crop up everywhere, the dynamics of the entire telecom industry start to change.

Traditional Telcos Will Have a Hard Time

Consumer VoIP and Voice as a Managed Service

For carriers targeting the consumer market, VoIP is inexpensive to run, and is already very competitive, with an estimated 400 providers in North America alone at the time this book was written. Just the other day, I saw a Google ad that read, "Start your own VoIP business for only $250." There are open source VoIP software solutions available, which means that any skilled programmer can create their own VoIP service, without the barriers of massive capital costs that have formerly limited competition to large manufacturers. It is very conceivable that the cost per telephone line will become ever lower as telcos become more efficient and cutthroat competition continues.

The Death of Long Distance

Already we see "all-you-can-eat" residential plans for long distance. It will happen in business too, inevitably. Some telcos in Canada are already offering free long distance within North America as part of a bundle in order to promote other services. Even rates to international locations are expressed in pennies per minute.

However, even though long distance pricing is approaching zero, organizations will continue needing to monitor usage to make sure that the calls being billed to them are actually resulting in a business benefit. All organizations are motivated to make sure that their employees are efficient at their work despite all the temptations. Similar to excessive Web surfing and personal emails being monitored and controlled, there will always be a need to watch the call detail records.

Vendor Consolidation

Back in the 1980's when the telcos' long-distance monopolies crumbled, a great many companies sprang up to offer cut-rate toll calls, built on the telcos' networks. Some of these resellers were founded with the sole aim of building a subscriber base and selling it to another carrier at a profit. Eventually, the dozens of smaller companies were consolidated into fewer players.

We are seeing a renewed continuation of that same trend today. Simply trying to determine which is the best offer for the organization, given the bewildering range of products and names for those products, has been a major challenge in the past. But as corporations demand simpler custom contracts, and newer technology like VoIP pushes pricing down even further, prices will continue to drop.

If virtually all voice and data transmissions become so inexpensive that it's not even worth the trouble to measure cents per minute anymore, or even gigabytes of transferred data, will telecom costs become so inexpensive that it almost becomes irrelevant?

What will happen to the massive telecom industry around the world as these newer technologies replace their older infrastructures, and erode their revenues? Telcos are getting in on the VoIP game too, in order to hang on to as much market share as possible. Will this just accelerate their decline in revenues?

Consider a large organization that spends $100 million per year today. This total consists mostly of data circuits, Internet access, mobile phones, local, and long distance service. Out of these, the only service that they will be able to charge much for is Internet access. A large enterprise such as a bank that has 1500 sites including ATM machines may only need to pay $40 per month for each, and maybe a few hundred or even thousand dollars for their larger head office and call center sites. This only adds up to $1 million to $2 million per year. Even if they pay $10 million instead of the $100 million they pay today, that still means that their vendors will eventually lose 90% of that revenue.

Considering that telco margins will continually become slimmer, the market will have to consolidate even more radically than it has to date, which means you will have far fewer vendors and invoices to deal with.

Chapter 16

The Rise and Fall of Telecom Management

What will the future of telecom management be like?

Telecom management as a profession has not always been as widespread. Several decades ago, before the proliferation of mega-corporations, and all the increased pace of mergers and acquisitions, there was a lot less need for professional telecom managers. Telecom management evolved on the data side from the computer scientists in white lab coats as networking began to be more commonplace and organizations needed dedicated staff to handle it. On the voice side, for many companies the need for telecom managers evolved from the switchboard operator, to the PBX technicians, and from more general facilities management.

Today, there is a tremendous need for organizations to have internal expertise on all the various new technologies, vendors, pricing options, bill auditing, and a whole host of other essential work. But will it always be this way?

Examining all the coming trends in this section, the one thing that is most obvious is that the massive shifts in the market and technology will change many aspects of telecom management. Not only will it be different, but there will be far fewer telecom managers. Over the next ten years or so, all these new technologies, and probably some others that aren't even on the radar today, will evolve into standards for sending data and placing phone calls.

What do all these future trends have in common? The purported benefits of each and every major development listed in this section of the book are to reduce and simplify pricing and operations. Sure, there will be the initial learning curve for each of these technologies, and trained professionals will be required in order to implement them. But as time goes on, and these go from being cutting edge to just ordinary, there will be a surplus of people with the skills needed to manage these technologies.

In ten years, if every employee in your organization has ultra high-speed wireless Internet access at their desktop and on the road, with every application including voice at their fingertips, when will they need to call the telecom department? They may need help fixing glitches with their connectivity, hardware, or software occasionally, but those might be handled by an outsourced IT department in another country. If long distance is virtually free, and companies pay a low fixed cost at each site for Internet access, how much billing needs to be examined?

What will become of every person who has dedicated his or her career to managing telecommunications? Sadly, many of them will find that their services are no longer required, and as corporate cutbacks continue, the bottom ranking staff will be let go.

This future is still a long way off, probably at least ten years from when this book is published, maybe even fifteen. In the meantime, it will be important for telecom managers to be as efficient as possible in their current environments, and learn as much as they can about these shifts in technology, especially VoIP.

On the bright side, there will always be a need for the individuals who bring the most value to the organization. In the aftermath of the dot-com bubble bursting, millions of software and related employees suddenly became unemployed, finding the new reality of the job market very difficult to cope with. On the other hand, the most talented and savvy of them continued on, and even those whose employers went bankrupt were quickly able to find new employment.

The same thing will happen in the telecom management space. Great telecom managers who are continuously looking to improve their knowledge and skills will find more opportunity available in the future, and will likely upgrade their salaries in the process. Since you're reading this book, you are likely one of those telecom professionals who is constantly seeking improvement.

Those who understand their organizations' strategic goals and are able to help in achieving those goals will always be in demand. Being able to drive more ROI is a talent that is always useful in any organization, and in any profession. If you are producing valuable information, and are demonstrating your department's ROI, you are then a very necessary part of the organization.

Special Free Gifts from the Author

As I was writing this book, I thought of changing the title to "What Great Telecom Managers *Do*" instead of "What Great Telecom Managers *Know*," because all the difference in the world comes from *doing*. To help you along, I have made some special arrangements for you to participate in these complimentary one-on-one sessions.

You will not be the victim of any sales pitch for consulting services or software. These telephone sessions are designed to give you the most impact for your personal career and your telecom department.

Gift #1: 30-Minute Session With a Business Coach ($120 value)

- How can you be sure that you get the financial rewards you deserve?
- Do you sometimes feel taken for granted?
- What can you do to get the recognition you deserve?
- What can you do to better demonstrate your ROI?

Together with business coach Willa Schecter, you will create a customized plan, and determine what would make the biggest impact to your job satisfaction.

You'll leave the call with a clear idea of what you can do to increase your profile within your organization and a plan of action that will help you get the most out of your career.

Willa is a graduate of The Coaches Training Institute, and

has completed her post-graduate accreditation. Willa sits on the Board of Directors for The International Coach Federation, Toronto Chapter.

Gift #2: 30-Minute Telecom and Wireless Cost Management Tactical Review ($240 value)

You will get insight on how you can reduce costs and improve the effectiveness in your telecom department. Address your toughest telecom and wireless cost management challenges, and get an outside perspective. If we feel that your circumstances are outside our area of expertise, we will decline the engagement and instead may recommend other resources to you.

To obtain your complimentary sessions: E-mail the information listed on the form below to bookgift@avemacorp.com, or make a copy of the form, fill it out and fax it to 416-364-5101.

Name _____

Title _____

Company _____

Phone _____

Email _____

Comments _____

Providing this information constitutes your permission for Avema Corporation to contact you regarding related information.

About the Author

Roger K. Yang is the founder and CEO of Avema Corporation. Since 1995, Avema has helped hundreds of organizations with their wireless and telecom costs, and Mr. Yang has been directly involved with most major clients. Recently, he introduced Avema's innovative 100% ROI guarantee.

Mr. Yang was a pioneer in telecom and wireless cost optimization software since 1997, and currently leads product development for Avema's telecom expense management software. This software powers the operations of firms around the world that deliver telecom expense management services to their own clients.

In addition to this book, he is also the author of several articles and information guides. Mr. Yang periodically conducts workshops for telecom managers on best practices of telecom environments and expense management.